FOLLOWING THE COAST

DUTCH COASTAL SHIPPING OVER THE YEARS

Willem H. Moojen

Following the coast

FOLLOWING THE COAST

DUTCH COASTAL SHIPPING OVER THE YEARS

Willem H. Moojen

The changes in types of coaster
1930 - 2005
Volume I

Lanasta

ISBN: 978-90-8616-034-1

Dutch edition: 'Van kust naar kust'
ISBN: 978-90-8616-029-7

NUGI: 969

1st print, October 2007

Translation: Elizabeth A. Molt
Merije Wennekers

Design: Lanasta
Photo editing: Dick de Vries

Print: Drukkerij Giethoorn Ten Brink,
Meppel (NL)

http://www.lanasta.com

Books by Willem H. Moojen published by Lanasta:

Dutch/English

ISBN 978-90-8616-004-4
- Nederlandse Koopvaardij in beeld 1950-1959 (I).

ISBN 978-90-8616-007-5
- Nederlandse Koopvaardij in beeld 1960-1969 (I).

ISBN 978-90-807822-4-2
- Koninklijke Marine in beeld: 1960-1969.

Dutch with English summary:

ISBN 978-90-8616-002-0
- Hollandsche Stoomboot Maatschappij. De vaart op het Verenigd Koninkrijk en West-Afrika. Deel I: Rederijgeschiedenis.

ISBN 978-90-8616-006-8
- Hollandsche Stoomboot Maatschappij. De vaart op het Verenigd Koninkrijk en West-Afrika. Deel II: De vloot.

Dutch:

ISBN 978-90-807822-9-7
- Midscheeps, De zeeman vertelt. 2e druk

ISBN 978-90-8616-020-4
- Vergezichten, Zilte poëzie met Leo Bersee.

Jacket photographs
Front: The ***Flinterzee*** *navigating the Westerscheldt on 26 September 2002.*

Back:
Above: The ***Duurswold*** *on trial cruise.*
Middle: The ***Willi Böhmer*** *unloading timber in Zaandam.*
Below: The ***Helene*** *enters IJmuiden on 11 October 2003.*

Following the coast

Foreword

In the last 75 years, Dutch coastal shipping largely has been vital to the image of our merchant marine. It started with peat shipping from the east of the Netherlands to the coasts of the former Zuiderzee (now IJsselmeer) and reached over 1,000 ships at its peak in the mid-fifties of twentieth century shipping. This book follows the ships from the 1930s until the present, and shows that the traditional coaster gave way to ships that became bigger and bigger. Consequently, the rather limited sailing itineraries could now cover a much wider radius.
It is not my intention to go deeply into this fascinating branch of shipping trade since literature about this subject is readily available. But like my series of '*Nederlandse Koopvaardij in beeld*' I propose to portray with photographs how the type of ships has changed throughout the years, and particularly in the 1950s, 1960s and the 1970s of the previous century when so many magnificent ships left the slips. The scale-up of the Dutch coasters has made the word 'coastal shipping' obsolete to a certain extent, as there are only a few coasters left that have permits for restricted navigation only. Coastal shipping today means 'short-haul trade' as almost all ships have unlimited navigation permits and swarm the globe.

This book shows the characteristic small coasters in days long past and the ships that we see nowadays. This review includes refrigerated ships, products and chemicals tankers, dry-load vessels, feeders and low-profile coasters. It is a tribute to Dutch shipbuilding and its results.
This is volume 1 of a series of books.
I hope that you enjoy it as much as I did when compiling this book.
Dutch coastal shipping, presently named short sea shipping, was and still is a household word.

Willem H. Moojen
Beverwijk, September 2007

A glimpse from the forecastle to the rear superstructure of the ***Atlantic River*** *during the loading of seed-potatoes in De Pijp harbour in Beverwijk on 8 December 2000.*

*Dutch coasters could be seen in English harbours almost daily. In the foreground the **Greta**, belonging to J. Schokkenbroek and completed in 1957, is here seen berthed in Whitstable in south-east England; behind her the **Solent**, belonging to G. Schoorl and completed in 1953.*
This is Dutch short-haul trade of the upper 1950s, which many shipowner, captain and crew will remember as the heydays of Dutch merchant navy.

Contents

Introduction

In the 1950s, and even earlier, we may have seen one or two coasters on the other side of the world, especially when we remember Van Nievelt, Goudriaan & Co's Stoomvaart Maatschappij N.V. from Rotterdam, which had coasters sailing a regular service to South America.
N.V. Stoomboot-Maatschappij Hilligersberg (Steamship Company Hilligersberg) in Amsterdam had built special coasters for a biweekly service, which soon became a weekly service between New York and Bermuda with the **Oleander** and the **Hibuscus**. The **Oleander** first sailed in bare-boat charter because the shipyard had built the ship at its own expense before selling her in 1956. In 1960, this line was extended with a biweekly service to Jacksonville as well as to Baltimore.
Koninklijke Nederlandsche Stoomboot Maatschappij N.V. (Royal Netherlands Steamship Company) in Amsterdam had a number of coasters sailing the West, deploying them as feeders. These coasters were part of the West-Indische Scheepvaart Maatschappij (West Indian Shipping Company) with Willemstad, Curaçao as their homeport.
Other liner trade companies or captains and owners would have their coasters also fan out to North or South America, but these were exceptions to the rule. Even in the Dutch Indies we saw coasters which had crossed the oceans to offer their services in the Indonesian Archipelago.

In former days

Of course, we have seen before that small Dutch ships already voyaged to far-away places and even to other continents but they were not engaged in those typical coastal shipping activities that we all still remember. These ships plied the seas with hoisted sails at the mercy of the wind: barges, smacks, spritsail tjalks with and without koff, galjoots, schooners, barques and frigates, heading for destinations like England, France and Germany and sailing up north to Scandinavia, Finland, the Baltic states and Russia. But countries along the Mediterranean, the Black Sea, the west coast of Africa, the Cape of Good Hope, Latin America, and even Cape Horn and Asia were no exception. Dutch sailors always had a real knack for trade and never lacked the courage to collect cargo from wherever they found it.

Extending the boundaries

In the 1960s, in addition to the almost traditional destinations like northwestern Europe, the Baltic States and countries around the Mediterranean, we saw more and more coasters daring to sail to other continents. Shipping companies and captains realised all too well that cargo and business opportunities were waiting for them at the other side of the Big Pond and even beyond. An other point that affected this trade was that on 1 June 1970 the 499 gross tonnage limit was abandoned and changed into a 75 metres limit, which was changed again into a 4,000 gross tonnage limit. This made it possible to build ships with a deadweight capacity of 9,000 tons. Nowadays, it is no longer the tonnage that is checked but the service area.

More and more shipping companies emerged, which extended their borders, including Dammers & Van der Heide's Scheepvaart- en Handelsbedrijf N.V. in Rotterdam, which began specialising in refrigerated shipping and which sent its ships to the Caribbean, North and South America, Canada and even to Australia. N.V. Scheepvaartkantoor Groningen, later renamed Seatrade Groningen, also built many fruit ships though. These specially designed and clean-lined ships could be seen not only in our Dutch waters but even far, far away. In fact, our coasters sailed all over the globe.

Many one-man shipowners extended their horizons too, like C. Nieuwpoort in Groningen, who owned the **Carpe Diem**, a fine coaster built in 1957. It sailed the Caribbean for months on end in 1959 and 1960, and frequented ports in Haiti, Jamaica, Puerto Rico and New Orleans in the U.S.A. Coastal shipping was transforming drastically, although there were enough coasters that stayed along the Dutch coast. And not all shipowners moved to far-away places because they ran regular services or called in at regular ports, like N.V. Wijnne & Barends Cargadoors- en Agenturenkantoren, Gebr Beck, E. Wagenborg's Scheepvaart & Expeditiebedrijf N.V., all of Delfzijl. By then, Wagenborg had become one of the larger shipping companies in our country, and it has been sailing worldwide for some time now. In Amsterdam resided the Vereenigd Cargadoorskantoor which was an umbrella for various shipping companies, the coasters of which would often be calling in at fixed ports, like Paris. N.V. Scheepvaartbedrijf 'Gruno', also in Amsterdam, sailed its own ships as well as ships in management mainly in European waters. Just like Comm. Venn. Terwogt en Lagers, also residing in the Dutch capital, which also functioned as a coordinator for several shipping companies.

*This is the **Carpe Diem** with her fine superstructure, here in Port of Spain. The awnings provide shelter from the heat.*

South America

Van Nievelt Goudriaan not only had ships for the long-haul trade in its fleet but also splendid ocean-going coasters which sailed various regular services. The fleet also included smaller units, some of which even were directed especially to South America to function as feeders on the Rio Paraná river for a period of time. The larger ships voyaged to harbours in Brazil and Argentina and to Montevideo, and smaller coasters took over their cargo from there. They sailed from the sandy Rio de La Plata estuary, with Buenos Aires on the other side of the river, far into Argentina's inland parts up till Asunción, the capital of Paraguay. It was adventurous sailing which nevertheless required solid seamanship. Despite the fact that often two pilots were on board, things could go wrong sometimes but this was usually due to local circumstances. No major accidents but usually minor incidents, like sailing into the banks because the river became so narrow that it looked more like a muddy ditch, etc. But then of course sailing in the jungle is quite different from sailing on the North Sea Canal, and even there something can go amiss.

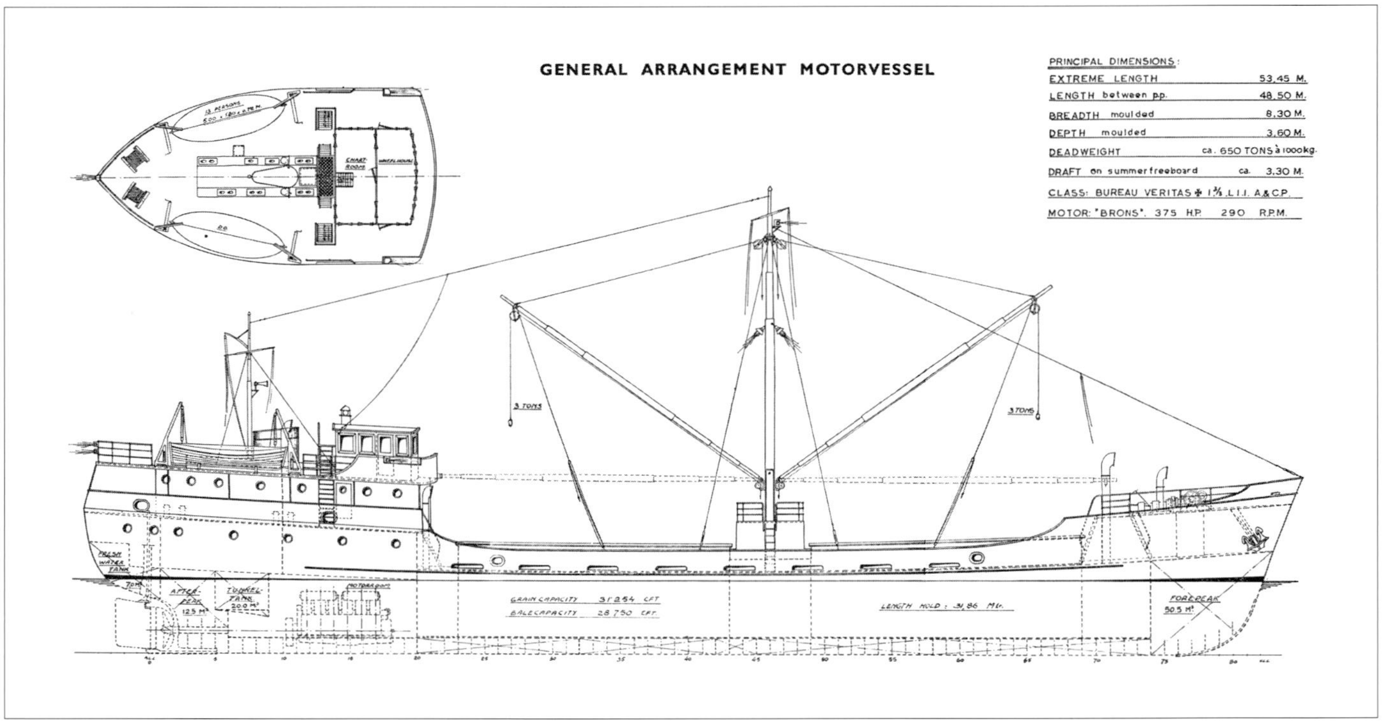

Africa

Coastal trading to North Africa and in particular to Morocco and the Levant already occurred in the 1940s and 1950s. Rotterdam shipping companies like Wm H. Müller & Co N.V., Dammers & Van der Heide's Scheepvaart- en Handelsbedrijf N.V., N.V. Reederij 'Nautiek' (W. Böhmer's Scheepvaartbedrijf) and N.V. Gebr. Van Uden's Scheepvaart- en Agentuur Maatschappij, to name a few, sent their coasters to North Africa. Müller ran a regular service to Casablanca with the **Walenburgh** the larger **Sandenburgh** and **Wickenburgh** and even ordered the building of two graceful ships, the **Domburgh** and the **Vrijburgh** specifically for this route. In fact, they were no longer coasters, but ships that fell under the class of home-trade or short-haul trade. They had accommodations for 6 passengers. Shipping company Dammers too regularly sent its ships to Casablanca and Kenitra. The Nautiek Company deployed amongst others the **Wiebold Böhmer** and the **Daje Böhmer** for its trade on Morocco. Van Uden sent its ships to Casablanca as well but also to Algiers and this company introduced its **Koningshaven**, **Merwehaven** and **Rijnhaven** in the Benelux Fruit Line as from December 1948. This line had Rotterdam as its homeport and called in at Antwerp and the Spanish east coast ports on her way to Casablanca. From 1952, the beautiful **Binnenhaven** sailed for the Frabenelux Fruit Line, which was established in 1951. This line had ships sail in service via Antwerp and Rouen to Casablanca, Port Lyautey (called Kenitra from 1956 onwards), Safi and Agadir. In fact, this line was the earlier Benelux Fruit Line. However, they now ran a weekly service and called in at more ports. More graceful coasters specially built for this service were to follow. Moreover, apart from these four shipping companies, of course many more coasters belonging to shipping companies and captains/owners sailed to North Africa. The N.V. Vereenigde Nederlandsche Scheepvaartmaatschappij from The Hague had the **Tayari** and the **Holland** in East Africa for feeder services.
The Koninklijke Nederlandsche Stoomboot Maatschappij N.V. (Royal Netherlands Steamship Company) sailed with As-type coasters, which were part of the holding Nieuwe Kustvaart Maatschappij, and with larger ships

A general plan of a type of coaster that was distinctive for the period of the late 1930s until the early 1950s. This type and its many variations have been decisive for the Dutch coaster fleet to expand and to become one of the largest fleets in the world.
In those days one could hardly find a port in or even outside Europe, without spotting the Dutch flag flying proudly from a coaster's aft mast.

that had limited passenger accommodation, to a great number of countries around the North Mediterranean and the Levant, North Africa and to ports in the Black Sea.

Ports and cargo

The real coastal trading, where ships would call in at smaller ports far inland, is almost over everywhere. Many harbours and ports are closed, filled in, or are used for inland shipping only. The supply of cargo has also changed. The times that amongst others much strawboard and especially coal were transported from traditional ports are gone, not in the least due to the closure of many mines. And when the ports closed, many industries moved away as well, taking with them the supply of cargo of which mixed cargo had always been a substantial part.
Naturally, mixed cargo is still being carried, the fact is that the vast majority is now mainly transported in containers, and from other locations. Timber is also still transported in large numbers, but now in packages and no longer plank by plank. In Zaandam we could see the coasters lying in rows unloading timber from the Scandinavian countries, and especially from Finland. Of course, timber is still being unloaded in Zaandam, but in a different way and at another location.
The clock cannot be turned back, and we ought to think ahead; while the industry adapts itself to the changing needs. But I, like so many others, still think back wistfully to those times when coasters could be admired well inland in towns like Alkmaar, Leiden, Utrecht, Venlo, Zwolle, Leeuwarden, Veendam, Middelburg and dozens of other towns.

Shipyards disappeared

Most coasters were built in the north of the country. In the provinces of Groningen, Friesland and Drenthe more than 55 shipyards could be counted in their heydays. Driving along the Winschoterdiep you could see them from afar, towering above everything. There were times that several coasters slid from the slips on one single day. Even this era now lies behind us. Many shipyards had to close their gates, as ships became larger and larger and the canals were not wide enough for them to navigate through. Other shipyards merged and tried to keep afloat, but when the classical coaster disappeared from the oceans, more shipyards followed. This called for a change of course and the Dutch shipbuilding industry started to specialise. The Dutch shipbuilding industry has always had a good reputation worldwide and this came in handy when more and more industrious companies buckled down to building more complex ships like gas tankers, product tankers, refrigerated ships and dredgers. Furthermore, we still are at the top in building very luxurious yachts. Now the tide is on the turn and although the future looks a lot better, the days of yore are now definitely over, now that low-wage countries have taken over. This is why some Dutch shipyards have subsidiaries or interests in these countries to keep some grip on the market.

Their round forms gave way to more austere styles

A Dutch coaster could easily be recognised by its architecture and many could single out Dutch ships from the ships from other countries. Take for example the small coaster from the mid-twenties to the mid-thirties with a capacity of about 200 to 300 tons, with its characteristic round back (cruiser stern) built at shipyard 'Gideon', vh J. Koster Hzn in Groningen. These coasters can be seen as the forerunners of the type of coaster from the 1950s that was to make our country so famous. There were of course many more shipyards building similar coasters in those days, such as N.V. Noord-Nederlandsche Scheepswerven in Groningen, Scheepswerf Gebr. Niestern & Co in Delfzijl, N.V. Scheepswerf 'Delfzijl' v/h Gebr. Sander, also in Delfzijl, and Scheepswerven Gebr. van Diepen in Waterhuizen, etc.

Coasters were often specially built in consultation with the owners for the cargo they had to transport. Thus we could see flush-deck and shelter-deck vessels, and vessels with a half shelter-deck or raised quarterdeck, coasters for the service to Paris and other low-profile coasters. But even these designs were changed again and throughout the years we see totally different types of coasters. Not only did the growth in scale play an important role but also the way in which the cargo was provided. As I stated before, much cargo is transported in containers, and the ships have been adapted accordingly. The ship's round forms and her sheer were replaced by austere lines and rectangular (box-shaped) holds. Efficiency first! The mostly smaller ships usually have one single hold and no more tween-decks, with the odd exception of course. Such a hold has the advantage that not only it is possible to place the containers in a better position, it also enables the transport of much larger pieces of cargo; think of project cargo which includes large machines and parts intended for the petrochemical industry. Did we use to see many coasters once that had been built to the liking of the owner or captain, which gave us the opportunity to see many, many beauties plying the waves, this too is something of the past. Nowadays, most 'coasters' - even the name has faded meanwhile - are often built in mass production to save costs. Some types of ships sail by the dozens, which slightly reminds us of the construction of the standard ships in the olden times.

The time that a lot of copper and teak were used in the decoration of the accommodation is also long gone. Partly due to safety requirements, but certainly also in an effort to reduce costs, the decoration of the cabins and other quarters on the ships has become much more basic and less labour-intensive to keep it all clean.

*The **Amyntas**, launched at E.J. Smit & Zoon's Scheepswerven N.V. in Westerbroek on 28 April 1955 and delivered on 15 July 1955, mainly sailed to countries around the Mediterranean and to the Levant. The coaster was owned by Nieuwe Kustvaart Maatschappij N.V. in Amsterdam, a subsidiary of KNSM.*

Changes in the type of coaster

The coasters from the 1930s were mainly of the flush-deck type, as these ships were very suitable for the transport of mixed cargo and timber. They would dominate the scene of our Dutch coastal trade for many years.
Because of the war, the 1940s were not exactly characteristic for one certain type of ship. As a matter of fact it produced many different types of ships. Naturally, all this had to do with ships of a certain class that were delivered by the yards in mass-production during World War II. Even the conversion of ships could lead to surprising results, as could be seen in the conversion of landing crafts. These years also saw a growing popularity of the raised quarterdeck type. This type of ship was very suitable for the transport of bulk cargo, such as coal and grain. However, we would also see many of these ships in the timber transport.
In the 1950s and 1960s flush-deck, shelter-deck and half shelter-deck coasters formed a majority. Here too, apart from carrying all sorts of other cargo flush-deck coasters were often deployed in the timber transport, while the shelter-deck vessels were mostly used for liner traffic. This type of coaster often had a 'tween-deck' which made it very suitable for the transport of perishable cargo like fruit and potatoes as well as large quantities of mixed cargo. But when trade was slow, these ships saw no harm in transporting China clay, which was not quite a popular load.
One type of coaster that entered the short-haul trade was the so-called Denmark coaster, also known as inside/outside ship or even irreverently called 'sand-scraper'. They were usually adapted inland waterways ships the gangway and hatch room coamings of which had been raised. And as the name already suggests: they often sailed to Denmark and its surroundings. One could say that these ships sailed on the same route as the flat-bottomed barges that scoured the countries along the Wadden Sea and the Sound, called Wad vessel and Sound vessel, though they often drifted further away. This type of ship, also known as Belt vessel, could already be seen in large numbers in the 1920s. The size of these small ships ranged between 125 and 250 tons.
Given that the Denmark coasters only sailed a limited radius and measured not more than 1,600 Gt, they not only had a certificate of tonnage for service at sea at their disposal but also an inland shipping tonnage certificate, in accordance with the 1969 International Treaty of London. Through the rivers they were able to reach places deep down in the hinterland, not only in the Netherlands but also in other countries in Europe. At this moment two of these typical Denmark coasters are still navigating our country.
One might even say that the low-profile coasters which appeared in the late 1970s were developed from these Denmark coasters type. Although these ships are much larger of course, even most modern ones can serve far into the hinterland due to their shallow draught and by lowering their superstructure.
To some extent, this has brought back the door-to-door transportation with small coasters of the olden days, although the smallest harbours, if they haven't been filled up or been made inaccessible, cannot be reached on account of their size anyway.
Nevertheless, in the 1940s and 1950s and even before, there were many ships to be seen reaching far into European waters of the hinterland, especially for trade to Paris. This was the so-called Paris liner, with her bridge usually amidships in the early days, but later also on the aft. These ships too had a small draught below the waterline and by lowering the

masts they were able to navigate the Seine under the many overhead bridges to reach Paris. They also passed many locks on their way. Several Dutch shipowners had scheduled services with these ships at that time and they had them built tailor-made for this purpose. The acme of these liners was reached in the 1950s, but then they went into decline, and they would call in only at Rouen.
An other type of ship is the ro-ro (roll-on/roll-off) ship. This is a type of ocean-going vessel which is mainly intended or used for the transportation of cargo that either entirely or partly can be rolled on board and rolled off again by using loading ramps specially equipped for this, which belong the the standard equipment of the ship.
A type of ship that could be seen more and more often on the oceans in the late 1960s and 1970s, with the odd exception, is the refrigerated ship, also called 'reefer'. Previously, they used to take coasters that were partly equipped with a cooled or refrigerated hold. Reefers are mainly deployed for the transport of cold-storage products like butter, meat and fish, but fruits and other perishable products are also carried along. And if there is not enough refrigerated cargo available they even carry mixed cargo or containers with them. It is impossible to imagine present-day ocean shipping without this type of ship.
The tanker is not new in coastal navigation either, but the safety requirements these ships have to meet are now far more stringent than in the 1940s and 1950s, for example when you look at the double hulled tankers in order to minimise the risk of calamities. When once there were tankers specially built or equipped for only one type of product, such as bitumen, we now have tankers that can transport bitumen, oil and chemicals in different cargo tanks at the same time. The multi-purpose container ship is a vessel that next to containers can take in mixed and bulk cargo as well and which is often equipped for heavy loads.
The heavy-load ship is primarily used for the transport of very heavy parts, like machines and cargo used in the petrochemical industry. These ships are specifically equipped with loading gear that can handle very heavy loads.
Another type of ship that has emerged in the last few years is the feeder ship, which is mainly used for transhipment of cargo from other vessels (large container ships that run a regular service) and to carry it to an other, usually scheduled destination. This principally concerns containers.
And the type of coaster that has actually replaced both the flush-deck and shelter-deck vessels nowadays is the single-deck coaster that can carry both mixed-, bulk- and project cargo and even containers. They are deployed in our merchant fleet in large numbers.
There are of course more ships built for specific purposes but they will be reviewed in a following volume.

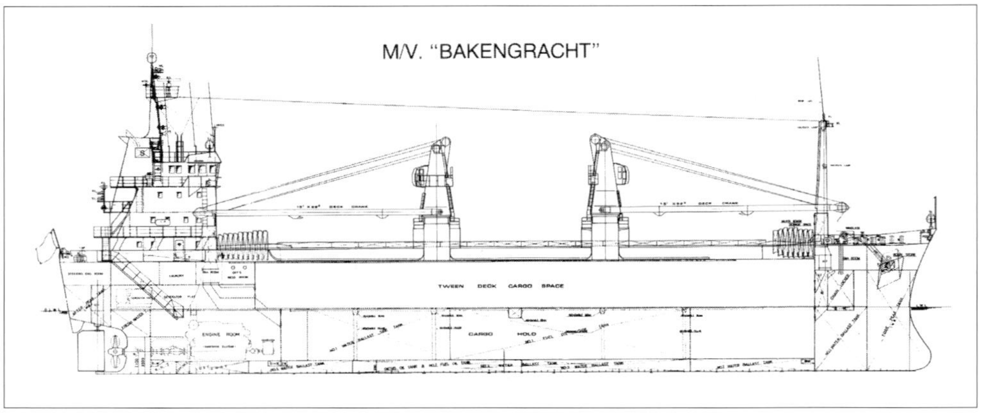

When compared to the general plan of the small coaster shown on page 10, it is clearly visible how the increase in scale of the ships has boomed enormously in the 25 years from date. In tonnage this means that a modern vessel like for instance the ***Bakengracht****, of which the general plan is shown here, roughly could take almost 10 times more cargo than an open shelter-deck ship and almost 18 times as much cargo than a closed shelter-deck vessel. This expansion would only increase even more in the years to come.*

The thirties

Flush deck **Democraat**, *in this photograph sailing the North Sea Canal, was launched with yard number 196 by G.J. van der Werff in Hoogezand on 3 March 1930. With her length of 33.75 metres and beam of 6.63 metres and a gross tonnage of 198 she was a small ship to carry 260 tonnes deadweight. A full complement consisted of 5 men. This little coaster was owned by Geert Vos from Hindeloopen and Freerk Krikke from Groningen when this photograph was taken.*

*The **Zeehond** was a stock example of Dutch coasters in the 1930s. She was built in 1930 by shipyard Scheepswerf Firma J. Smit & Zoon in Foxhol for Willem Nooitgedacht in Groningen. She sailed until 1955, and was replaced on 23 November 1956 when Mr. Nooitgedacht put another **Zeehond** from March 1937 into service.*

*Here we see the in 1931 built **Castor**, belonging to Jurrien Kruidhof and Jilke Hermannus Bos, on the North Sea Canal on her way to Amsterdam. She had the odd-looking covered bin on the flush deck that was often used on fishing vessels. The rather large davits on the stern replaced the single davit on the ship.*

Johannes Muller from Dordrecht had the 300 tons ***Mulan*** *built by N.V. E.J. Smit & Zoon's Scheepswerven in Westerbroek in 1931. The coaster was barely 37.50 metres long and had a 3-cylinder 4S. C.S.A. Stork diesel engine of 150 bhp on board to ensure a speed of 8 knots.* ***Mulan*** *stands for the names of Johannes Muller and his wife Kommertje Langbroek. He transferred the ship to his son Jan Muller in December 1937.*

*The **Ador**, in the photograph above sailing along the English south coast and below on the New Waterway in 1967, was the former **Eban** delivered by N.V. Scheepswerf 'Vooruitgang', Gebr. Suurmeijer in Foxhol on 3 September 1931. Initially, the flush deck coaster belonged to S. Rusthoven and M. Puister from Groningen, but in March 1963 Mr Puister was the only owner of the ship. She had a gross tonnage of 288 and transported a lot of timber and coal.*

The flush deck coaster ***Maartje*** *had Middelharnis as her homeport and was owned by Adrianus Goumare from Hellevoetsluis. She was built by N.V. Werf Vooruit in Enkhuizen in 1932, and measured 40.27 metres length with a gross tonnage of 348, and a net tonnage of 234 with a carrying weight of 400 tons. In 1962, the* ***Maartje*** *was sold to Paul Lindqvist & Co. (managed by O/Y Lars Krogius A/B), in Porvoo/Borgå, Finland, and rechristened the* ***Svanvik****.*

Below: The ***Maartje****, here with a grey hull, is seen here loaded up to her watermark and berthed near Suttons'Coal in a typical English harbour.*

Following the coast

*Left page: A hard worker putting in at English port in about 1952. This 230-tonne flush deck **Mado** was completed on 26 July 1932, and transferred to Douwe Stienstra in Groningen. The coaster was fitted with a 3-cylinder 4S. C.S.A. Klöckner-Humboldt-Deutz diesel engine of 150 bhp fetching 8 knots.*

*The photograph above shows the **Dagny** in an English harbour. On 4 January 1950, Cornelis Timmer from Groningen bought the coaster from Pieter de Wit from Groningen, who had operated the ship as the **Cateli** from 14 June 1932. In turn, Hendrik Groen from Groningen bought this small coaster from Timmer on 21 July 1953 and named her **Donata.** The photograph shows the **Donata** in IJmuiden with the former accommodation of Radio Scheveningen in the background. The flush deck coaster had a length of 32.62 metres and a breadth of 6.53 meters, with a carrying capacity of 230 tons.*

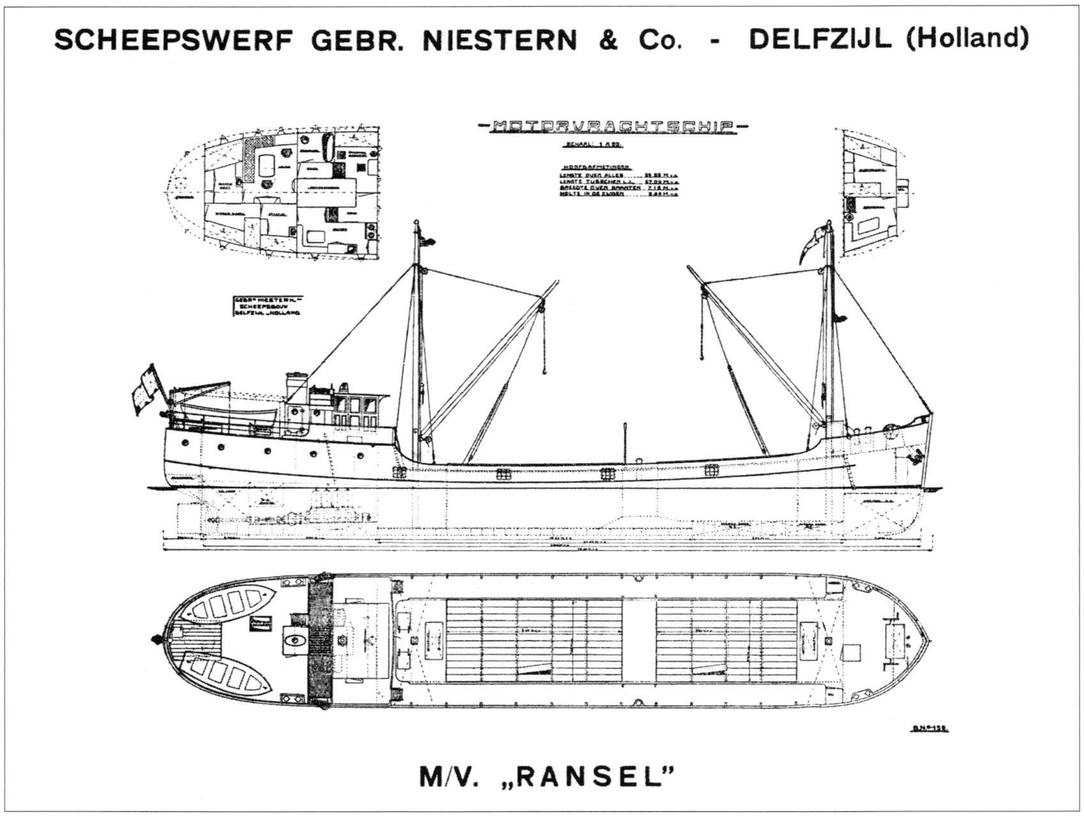

The flush deck coaster ***Ransel*** *was built with number 192 by Gebr. Niestern & Co. in Delfzijl. She was launched on 24 May 1933, and transferred to captain and owner Jannes Teerling from Delfzijl on 29 June 1933. She had a length of 37.72 metres and a beam of 7.19 metres.*
This small coaster had a gross tonnage of 258 and a carrying weight of 353 tons. In May 1935, the ***Ransel*** *was converted at the building yard and extended to 43.76 metres which increased her carrying capacity to 460 tons. The* ***Ransel*** *was sold again on 16 November 1937, because captain Teerling ordered a larger coaster that was put into commission as the* ***Ransel*** *on 8 September 1938. Roelof Buisman from Zwartsluis and Jetze Fokke de Groot and Hermanus van den Broek from Rotterdam became the new owners, and rechristened her* ***Servus****.*

The ***Ransel*** *under construction at the wharf.*

Her status is near completion.

Back in Delfzijl after her trial cruise.

Flush deck coaster ***Egbert-S*** *berthed in IJmuiden. She was constructed in 1933 and belonged to shipping company Rederij 'Egbert-S' of Lambertus, Jan & Jans Schut from Groningen, who bought the 37.40 metres coaster from Geert Leinenga from Delfzijl on 16 July 1956, who sailed her by the name of* ***Gersom*** *for over two years. Before that, she sailed with the names of* ***Kathleen*** *and* ***Thriantha****..*

*This is the **Vesta** docked in Groningen. She was a product of N.V. Scheepswerf 'Delfzijl' v/h Gebr. Sander from Delfzijl, and transferred to Klaas Beck in Groningen on 4 June 1934.*
This flush-decked coaster with a gross tonnage of 200 was confiscated by the Germans on 10 October 1940, but found back in Hamburg in May 1945, and returned to her rightful owner after repairs.
*The photograph below shows the ship again in an English harbour, somewhat altered and renamed the **Sont**.*
On 6 August 1955, Klaas de Vries from Ruischerbrug near Rotterdam became the owner.

This is the ***Birmingham*** *in 1952. She was delivered on 17 November 1934 by Rotterdamsche Droogdok Maatschappij N.V. to the Internationale Kustvaart Maatschappij N.V. in Rotterdam. The flush deck coaster had a length of 49.20 metres and a breadth of 7.69 metres, with a 8-cylinder 2S. C.S.A. Werkspoor diesel with 500 brake horsepower.*
The ***Birmingham*** *was a regular customer in English harbours.*

Here we see the ***Dubhe*** *arriving in an English harbour. She was from 1935 and belonged to Hendrik Jan Engelsman & Pieter Smeltekop from Delfzijl. The flush deck vessel measured 265 tonnes and had sailed for the two owners from 16 March 1955, but was sold with the same name to Wilhelm Poppe in Varel, Germany on 22 May 1963 to serve as a mussel dredger.*

*The **Rex** was built in 1935 by N.V. Noord-Nederlandsche Scheepswerven in Groningen. This 400-tonne flush-deck type of coaster measured 40.61metres in length and 6.97 metres in breadth and was owned by Stoffer Rusthoven from Appingedam. A full complement consisted of 7 hands.*

*The photograph above shows the **Rex** on the New Waterway in 1967; the photograph below in English waters on 11 September 1958.*

*Here is the **Zijpe** built by N.V. Corn. Appelo's Scheepswerf. Electr. Scheepsbouw- en Reparatiewerf in Zwartsluis in 1936, with Wolfert Otte from Bruinisse as her shipping company.*
This small coaster with call sign PIZV, mainly sailed to and from the countries along the Baltic, the North Sea, the English Channel and the Irish Sea. The flush deck vessel had a length of 34.14 metres and a breadth of 6.29 metres, with a gross tonnage of 198 and a cargo capacity of 225 tons.

*In the next page, the **Zijpe** is seen as the **Adriana** in Rotterdam port, with Jan Erhart Fransen from Ruinen as the owner.*

The photograph above shows the ***Prinses Margriet*** *from 1961 of the Oranje Lijn (Maatschappij Zeetransport) N.V. in Rotterdam in the background. Also spotted is the weather ship* ***Cumulus*** *from 1944 moored at Parkquai, belonging to the Dutch Civil Aviation Authority in The Haque.*

The photograph below shows the ***Adriana*** *crossing the* ***Batavier V*** *from 1959, owned by Wm H. Müller & Co. N.V. from Rotterdam.*

A beautiful snap of the trial run of the ***Duurswold****, built in 1936 at Firma Gebr. Niestern & Co. in Delfzijl for the captains and owners Alderd Lenting and H.G. Wijngaard. The flush deck coaster had a length of 37.25 metres and a breadth of 6.55 metres.*
This 255-tonne coaster sailed as far as north Africa, the Cape Verde Islands, the Azores, the Mediterranean and the Red Sea till Perim on the south coast of Yemen.

*One of the few photographs taken of the **Nieuwaal**, bought by Oranje Lijn (Maatschappij Zeetransport) N.V. in Rotterdam on 27 March 1940. The 462-tonne coaster was built in 1936, and made some voyages to Reykjavik to transport frozen fish for N.V. Hollandsche Stoomboot Maatschappij in Amsterdam, after World War II.*

*The **Casana** was launched as the **Hada** for Johannes Sint in Dordrecht by N.V. Scheepswerf 'Gideon' v/h J. Koster Hzn in Groningen on 5 December 1936, and was taken over by Schellen Scheepvaart & Bevrachting N.V. in Rotterdam in 1947. The **Hada** was renamed **Casana** by the Schellen company in 1949. The shelter deck coaster measured 382 tonnes and sailed also to Paris.*

The photograph above shows the ship in her old form in the 1950s, and the photograph below shows her beautified remake.

*Willem Nooitgedacht from Groningen had been the owner of the **Zeehond** since 23 November 1956. The flush deck coaster was launched by E.J. Smit & Zoon's Scheepswerven N.V. in Westerbroek on 3 February 1937. The **Zeehond** was 37.88 metres long with a breadth of 6.64 metres. He sold the 285-tonne ship to A. Hansen in Åbenrå in Denmark on 13 April 1962, who christened her the **Kongsmark**.*

Shipyard N.V. C. van der Giessen & Zoon's from Krimpen aan den IJssel had six coasters of this type built between 1936 and 1939, with some differences in their superstructures and their masts.

The upper photograph shows us the ***Haskerland*** *of Scheepvaart en Steenkolen Maatschappij N.V. in Rotterdam on the North Sea Canal near Buitenhuizen; she was completed on 13 March 1937. The photograph below shows us the 581-tonne coaster in English waters.*

*Her name **Agiena** proudly displayed in large letters on her bows. This flush deck coaster was completed by N.V. Noord-Nederlandsche Scheepswerven in Groningen on 20 March 1937. She had a length of 41.10 metres, a breadth of 7.01 metres with a gross tonnage of 336 and a carrying capacity of 415 tons.*

The photographs show the ship almost jam-packed with a load of China clay in Charlestown in southwest England. The owner was captain Hendrik Evert Tattje from Groningen; B. Oeseburg was her co-owner from 1951.

AGIENA

TON-S
ROTTERDAM

TON-S

*The upper photograph shows the 650-tonne **Ton-S** on her trial cruise on the New Waterway near Rotterdam on 30 June 1937. The photograph on the left shows the coaster on the North Sea Canal with a full load of oil barrels of Van Leer Vatenfabriek. She was owned by N.V. Zeevaart Maatschappij 'Ton' and transferred to N.V. Rotterdamsche Kustvaart Centrale in 1946.*

*Here on the right she is seen with a grey hull and somewhat altered. The **Ton-S** was sold to Rino Camalich in Livorno/Leghorn, Italy, in 1960, and rechristened **Dinorah Camalich**.*

The ***Audacia*** *was transferred to Johannes Rilistinus Salomons in Stadskanaal as the* ***Insulinde*** *on 8 May 1937. The ship was sold to Olie Scheepvaart & Bevrachting Mij in Rotterdam in 1940, and rechristened the* ***Audacia****. She was confiscated by the Germans on 25 July 1941, and returned to Salomons as the* ***Insulinde*** *after the war. In 1946, Firma C. Minnaar in Overschie became the owner and gave her back her original name. In 1950, C.V. Exploitatie van het m.s. 'Timor' became the owner, renamed her '****Timor****', and gave her in management of J. Salomons from Rotterdam.*

The upper photograph shows her as the ***Audacia*** *in Rotterdam; below as the* ***Timor*** *entering an English harbour to pick up a load of pitch on 12 November 1950.*

*The **Plato** seen above at Levant Quay in Amsterdam, and (left) on the North Sea Canal. She was owned by West-Indische Scheepvaart Maatschappij (KNSM), which had its head office in Willemstad on Curacao. This coaster had a length of 57.45 metres and a capacity of 583 tonnes and was one of a series of nine almost identical ships. The **Plato** had accommodation for 11 crew, and was completed by N.V. Scheepsbouwwerf Gebroeders Pot in Bolnes on 18 August 1937*

*After her trial cruise on the Ems on 1 and 2 September 1937, the **Celebes** was transferred to captain and owner Roelof Jan Kajuiter from Groningen by E.J. Smit & Zoon's Scheepswerven N.V. in Westerbroek. After the German invasion of the Netherlands, the **Celebes** diverted to England where she sailed for the Allied Forces and aided coastal convoys.*
*On 30 March 1941, she departed from Liverpool in the direction of Falmouth and was reported missing. To this day, no one knows what has happened with the ship and her full complement of seven crew. Had the **Celebes** foundered by acts of war or had she struck a mine? We may never know.*

*The flush-decked **Wega**, the former **Deo Duce** owned by Johannes Boll from Winsum, was with her 47.50 metres suitable for trade with Paris. The ship was built with yard number 1101 by D. & Joh. Boot N.V. from Alphen aan den Rijn in 1937, and sailed from 1955 for Rederij Express in Groningen under the management of Mello Wedema and others.*

The upper photograph was taken in Jersey, and in the photograph below she was leaving Jersey.

This is the ***Patria****, transferred by N.V. Scheepswerf 'Foxhol' v/h Gebr. Muller in Foxhol to Hendrikus Groen in Groningen on 13 January 1938. The* ***Patria*** *was 43.44 metres long with a beam of 7.55 metres and had a raised quarterdeck, with a gross tonnage of 343 and 410 tonnes deadweight all told.*

This photograph shows the ***Patria*** *on the North Sea Canal in charter of the N.V. Hollandsche Stoomboot Maatschappij, Amsterdam.*

E. Wagenborg's Scheepvaart en Expeditiebedrijf N.V. in Delfzijl took over the ***Weltevreden 7*** *from Houthandel v/h. de Wed. M. Boogaerdt & Zoon N.V. in Krimpen aan den Lek, but let the ship keep her name. This happened on 14 March 1953. She measured a length of 37.55 metres and a width of 6.55 metres and was of the flush-deck type. On 23 December 1953, she was renamed* ***Hunzeborg****. On 1 October 1959, the* ***Hunzeborg*** *was sold to Danish shipping company Rederi I/S Baltic (H. Rasmussen) in Fakse Ladeplads, she was not renamed.*
The upper photograph shows the ***Hunzeborg*** *sailing under Dutch flag, and the lower photograph with the Danish flag flying on the aft.*

*The **Walenburgh** from 1938 belonging to Wm H. Müller & Co N.V. in Rotterdam serviced also the Rotterdam-Bordeaux-Casablanca-London and Rotterdam Line.*

*The photograph left shows the **Walenburgh** (a raised quarterdeck type) leaving an English port in the 1950s. There is a glimpse of the **Albert D** (1949) on the left.*
Below: Sailing in the English Channel.

*The photograph left shows the **Aardenburgh** entering Rotterdam port. She was almost identical to the **Walenburgh** and was transferred as the **Scheldt** to English daughter Vianda Steamships Co. Ltd., in July 1938. Then she was transferred to Müller in April 1955, and rechristened **Aardenburgh**.*

*This is the **Beja** belonging to Eeltje Roelfs Prenger from Deventer in an English harbour. She was built by Bodewes Scheepswerven in Martenshoek where she was launched on 19 October 1938 as the **Omlandia** for Geertje Albers-Panjer from Hoogezand. She sailed by the names of **Looiersgracht** and **Mutatie**, and Mr. Prenger bought the shelter deck coaster on 20 April 1956. The ship was powered by a 6-cylinder 4S. C.S.A. Humboldt-Deutz diesel engine with 300 bhp to sail at 9 knots.*

*Before Arend Jan Verbrugge from Portugaal bought the **Aiglon** of 1938 on 8 February 1963, she sailed by the names of **Barcarola**, **Martha** and **Ceram**. She partook as **Martha** in the landing operations in Normandy in July 1944. The ship had a raised quarterdeck with a gross tonnage of 336 and a carrying weight of 425 tonnes. The **Aiglon** was powered by a 2-cylinder 2S. C.S.A. Brons diesel engine with 224 bhp doing 8.5 knots.*

*A unique photograph of the **Dollard**, taken on the Westerscheldt during one of the few voyages that she was to take under Dutch flag. Construction was completed by Scheepswerven Gebr. van Diepen N.V. from Waterhuizen for Poppo Dekker from Groningen on 3 March 1939. On 14 September 1939, this fine coaster was sold to Rederei A/B Staffan in Gevle, Sweden and rechristened **Sigvard**.*

*Gerrit Brijder from Kampen became the owner of the **Nezo** on 14 July 1939. On 16 January 1953, he sold the ship to E. Wagenborg's Scheepvaart en Expeditiebedrijf in Delfzijl. The flush deck coaster measured 270 tons with a length of 37.55 metres and Wagenborg rechristened the coaster **Delfborg** on 22 September 1954. On 19 October 1959 they sold her to Fru Tessa Möller in Odense, Denmark, who renamed her **Adi**.*

She was launched as the ***Heiny*** *for Fokke Johannes Groot from Kampen on 21 June 1939, and bought by Wm H. Müller & Co N.V. from Rotterdam on 23 April 1940 and renamed* ***Nijenburgh****. This coaster measured a length of 44.45 metres and a breadth of 7.97 metres with a capacity of 530 tonnes. After being requisitioned she sailed for the Germans throughout World War II. On 16 March 1946, the* ***Nijenburgh*** *returned to Rotterdam. She used to sail the trade route to England.*

The ***Narwal*** *standing clear of the water in Par on 2 July 1964. The ship was owned by Pieter van Streun from Delfzijl and was completed by N.V. Scheepswerf 'Vooruitgang' in Foxhol on 11 August 1939. She was a flush-deck type and had a length of 37.86 metres and a breadth of 7.55 meters with a tonnage of 330. She was sold by the same name to Bror A.I. Karlsson in Kalmar, Sweden, in September 1967.*

Right page: The ***Jenjo****, here probably moored near an English harbour. Messrs H.J. v.d. Meer and A. Koster from Harlingen were the owners of this 270-tonne flush deck coaster from 10 May 1955. The complement of this coaster consisted of five men and was completed as the* ***Jozo*** *by shipyard Scheeps-werf Gebr. Niestern & Co. in Delfzijl for Ruurt Fekkes in Groningen on 31 August 1939.*

JENJO
JENJO

*Thijs Jonker from Groningen bought the **Alberta** (1939) from Jacob Rusthoven in Groningen in 1955, who had her in service as the **Capella.** In 1961, he extended the flush deck coaster with almost 8 metres at shipyard Scheepswerf Gebr. Niestern N.V. in Delfzijl to increase her carrying capacity from 314 to 375 tons.*

Upper photograph: Before lengthening.
Lower photograph: After lengthening.

The Forties

*On 5 June 1940, the **Batavier** was completed and transferred to Jan Muthert from Groningen. The coaster was originally fitted with two rolling cranes, which were removed later as they fell short of their purpose. The coaster was commandeered by the Germans on 29 April 1941 and returned to Delfzijl on 26 January 1946 where it was rebuilt by N.V. Scheepswerf 'Delfzijl' v/h Gebr. Sander. On 19 July 1947, E. Wagenborg's Scheepvaart en Expeditie-bedrijf and Firma Gebr. Niestern & Co, both from Delfzijl, took an interest of 1/3 part in the ship.*

*The photograph above shows the **Batavier** on the North Sea Canal, and below in English waters.*

On 17 August 1940, she was completed by Gebr. G.& H. Bodewes from Martenshoek as the ***Uranus*** *and transferred to A. Goudriaan in Rotterdam. The same month she was transferred to N.V. Reederij 'Muron' from Rotterdam, a member of the Hugo Stinnes A.G. group from Hamburg. She sailed in service of the Germans from December 1940, and under the management of Hugo Stinnes Reederei A.G. she was renamed* ***HAST-I*** *on 9 August 1941. After her return to the Netherlands in 1946 she was in commission again in May 1947. The unusual name of HAST was short for Handelsagenturen Scheepvaart en Transport (Trade Agencies for Shipping and Transport).*

These photographs show the ***HAST-I*** *in English waters. Mark the two different davits.*

m.s. "EEMSTROOM"

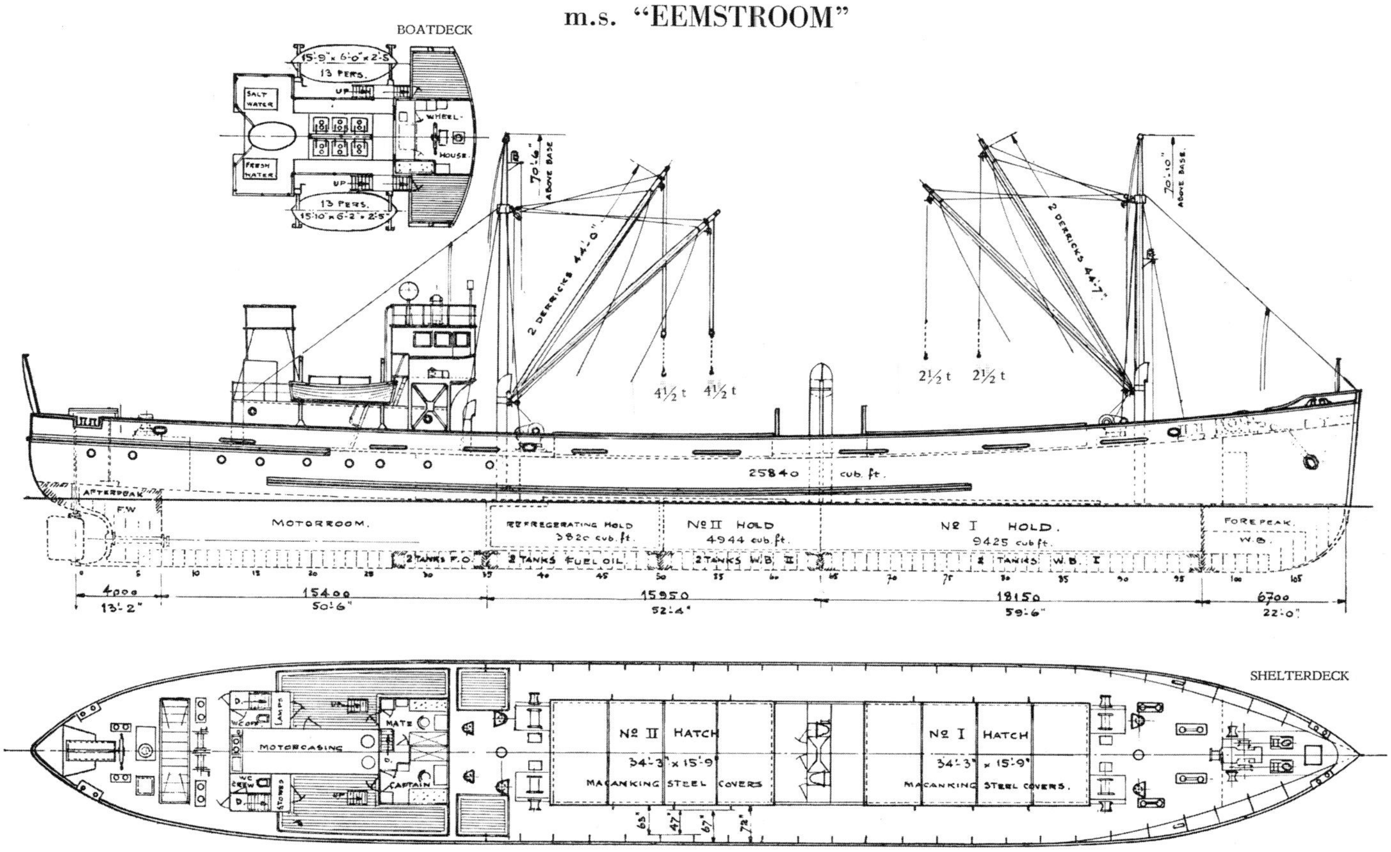

*The final touches to the 544-tonne open shelter-deck coaster **Eemstroom** was given by N.V. E.J. Smit & Zoon's Scheepswerven in Westerbroek on 9 July 1941, and she was soon to be transferred to N.V. Hollandsche Stoomboot Maatschappij in Amsterdam. However, she was requisitioned by German occupying forces, but was transferred in Hamburg to the Dutch authorities after the war in May 1945. After repairs at Nederlandse Dok Maatschappij N.V. in Amsterdam she sailed again as the **Eemstroom**.*
*On 4 November 1963, she was sold to J. Schokkenbroek in Groningen and renamed **Accres**.*

*Left photograph: The **Eemstroom** in Amsterdam harbour. Right photograph: As the **Accres**.*

*The **Blue Boy**, built in 1942 and belonging to N.V. Javo Scheepvaartbedrijf (Jac. A. Vonk's Handelsonderneming C.V.) in Rotterdam, was a LCG(M) type ship, which stands for Landing Craft, Gun (Medium).*
*The ex-**LCG(M) 123** of the Royal Navy was renovated by N.V. Holland-Nautic Scheepswerf en Machinefabriek in Haarlem and was into commission in 1947.*
The coaster was powered by two 6-cylinder 4S. C.S.A. Stork-Ricardo diesel engines of 240 bhp each and two propellers.

*N.V. Vrachtvaart Maatschappij 'Trako' from Rotterdam was the owner of the former English landing craft **LCF 23** (1943). After being converted at Dok- en Werf Maatschappij Wilton-Fijenoord in Schiedam and given the finishing touches by Firma Gebr. Bezemer in Dordrecht, she could start her trial cruise on the New Waterway near Rotterdam on 13 August 1949. She was named the **Lacra**, which is short for Landing Craft. The certificate of tonnage listed a gross tonnage of 497,41 and a deadweight of 750 tons, but it proved in practice that the volume of the ship was much larger and actually measured a gross tonnage of 524 and a deadweight of 1,296 tons. This made her the largest coaster in the Netherlands and she was to top the list for years to come.*
Other coasters claimed the title as well, but were so unfortunate not to match up.

*N.V. Teerbedrijf Uithoorn in Uithoorn bought a Chant-type coaster tanker from Sweden on 17 February 1951. After conversion, the small tanker was put into service as the **Gemma** (1944) on 2 April 1951. During a trip from Bilboa to Londen with a cargo of 300 tons of sulphuric acid in heavy sea her engine failed. She capsized and stranded between St. Jean de Luz and Guethary near Senix in southwest France on 29 December 1951, with a sad loss of 7 lives.*

*Left: The converted English Admiralty trawler of the 'Isles' type was an odd one out in the Dutch merchant fleet. Launched by Cook, Welton & Gemmell Ltd., in Beverley on 30 July 1942, she was commissioned as a buoy ship for the Royal Navy as the **Ulva T 248**. After being sold to Norway she was converted into a coaster. Roelof Boerma from Groningen bought the coaster in June 1950 and named her the **Suriname**. The ship did not sail long for Boerma, as he sold her again in August 1951. In September 1951, he bought a second **Suriname**, which was the former **Mies** from 1933. This coaster was sold in December 1953.*

The N.V. Koninklijke Paketvaart-Maatschappij from Amsterdam (Royal Packet Navigation Company) was assumed to have 20 ships in its fleet of the U.S. Army Coaster FS (Freight and Supply) type. All ships were built by Albina Engine & Machine Works in Portland, Oregon, USA, where the Dutch colonial administration of the Dutch East India islands, which was established in New York at the time, took over an annulled building contract from the shipyard. These ships were known as the Sitsen class ships, and were owned by the Dutch-Indonesian government and managed by KPM in Batavia (Indonesia), who was to take over these ships.

The photograph shows us the ***Sitsen*** *on trial cruise on 20 September 1945. She is still in grey, which was a common colour during the war.*

*The upper photograph shows us the **Labuan Hadji** of the same class, here loading copra at the Cocos Islands near Australia in May 1962. This little coaster sailed earlier with the names of **Tarakan** and **Bosnik** before the name **Labuan Hadji** was given to her in 1951.*

Below we see the coaster unloading wine casks in Dili Bay (Port Timor) in May 1962. The coaster was completed on 25 February 1946, and measured 54.86 metres in length with a beam of 8.86 metres and had a carrying weight of 586 tons. The full complement of crew was not less than 28 and there was accommodation for three to four passengers, with 49 deck passengers.

*The **Tyro** from 1946 belonged to N.V. Motorboot Maatschappij 'Bynia' from Harlingen, a full daughter company of N.V. Maatschappij Zeevaart, Rotterdam. She had a deadweight of 647 tons and a raised quarterdeck.*

*The upper photograph shows the coaster in Swansea (probably) in the mid-fifties; the photograph below was taken while navigating the North Sea Canal. She sailed the trade route to England or serviced the Holland-Ireland Line. The **Tyro** was sold to Ditta Giuseppe Dormio in Bari, Italy, in 1961, and rechristened **Bucador**.*

Though built in 1946, the ***Grebbestroom*** *looked very much like the standard ships mass-produced in England during World War II. She was launched at Goole Shipbuilding & Repairing Co. Ltd., in Goole on 29 July 1946, and measured a length of 71.44 metres and a tonnage of 1,228. She was transferred to N.V. Hollandsche Stoomboot Maatschappij in Amsterdam on 31 October 1946. The company deployed the* ***Grebbestroom*** *for the London and the Mersey lines. In 1951, the ship was fitted with a refrigerated hold. The open shelter deck vessel had two holds, six derricks and a heavy 15-tonne boom, and accommodation for two or three passengers.*

Upper photo: The ***Grebbestroom*** *navigating the North Sea Canal, the two other photos in English waters.*

*Flush-deck coaster **Bill-S** was launched by N.V. Scheepswerven Gebr. van Diepen in Waterhuizen on 21 May 1947, and transferred to N.V. Zeevaart Maatschappij 'Bill-S' early August.*
She was managed by N.V. Rotterdamsche Kustvaart Centrale.
*In 1959 the coaster with a carrying weight of 650 tons and a length of 55.67 metres was sold to B.E. Holmstrom in Gothenborg and renamed **Meridian**.*

*The photograph above shows the **Bill-S** in English waters; below on the North Sea Canal.*

*Page right: Hendrik Oldenburger and Mrs R. Buining-Siersema from Groningen were the owners of the **President Roosevelt**. This coaster was from 1948 and had a raised quarterdeck and a carrying capacity of 560 tons. In the photograph we see the ship leave an English port in the 1950s.*

PRES.ROOSEVELT
GRONINGEN

Both photographs show the same coaster (with a gross tonnage of 278) in an English harbour; above as the ***Arent. M*** *of N.V. Noord-Nederlandse Kustvaart in Groningen, and below as the* ***Adara*** *when taken over by Arend van de Veen from Groningen on 16 February 1953. The coaster was built in 1948 with a length of 42 metres and a carrying capacity of 350 tons and had a certificate for transporting timber.*

Page right: The upper photograph shows the ***Henrica*** *after being transferred to Rotterdamsche Zee- en Kustvaartreederij (RKC) on 17 June 1948. She had a raised quarterdeck with a length of 52.51 metres and a carrying weight of 680 tons and was constructed by E.J. Smit & Zoon's Scheepswerven N.V. in Westerbroek.*

The photograph below shows the same ***Henrica*** *after leaving an English harbour. She was transferred to N.V. Rotterdamsche Kolen Centrale in March 1952. The open railing next to the hatchway coaming was then closed from the gangway.*

HENRICA

HENRICA

The flush-deck coaster ***Berend-N*** *was owned by the brothers Jan, Berend and Kornelis Niestern, who incorporated the small vessel in the N.V. Martenshoek in Delfzijl. She was 35.59 metres long with a beam of 6.55 metres.*
There was a 4-cylinder 4S. C.S.A. Deutsche Werke AG MaK diesel in the engine room, good for 195 bhp. She was sold to Prospero Baiz & Co in Willemstad, Curacao, in 1961 and renamed ***Hertha****. She is seen here on the Ems in one of the rare photographs that were taken of the ship after she was sold.*

Below: This coaster was launched by N.V. Scheepswerf 'Vooruitgang' Gebr. Suurmeijer in Foxhol and christened the ***Carpo*** *by Mrs Van Zalm-Van Eldik. The coaster was owned by N.V. Maatschappij Zeevaart and in management of Hudig & Veder N.V. in Rotterdam, but she sailed only shortly for her shipping company. The* ***Carpo****, in this photograph navigating the North Sea Canal, carried a load of coals for a trip from Swansea to Amsterdam when the storm broke at 17 miles southwest of Lizard (Cornwall) on 27 November 1954. She sank, and 12 crew perished.*

Jan Smit Czn. N.V.'s wharf in Alblasserdam signed for the ***Batavier I****. She was launched on 10 November 1948 as the* ***Aardenburgh****, but was renamed before being put in commission on 26 February 1949. The shelter deck vessel had a length of 61.80 metres and a beam of 9.64 metres, and was deployed for the Batavier Line to London by Wm H. Müller & Co. N.V.*
From 1961, she sailed for the Bacoro Lijnen in partnership with Cornelder's Scheepvaart Maatschappij N.V. and Hudig & Pieters Algemene Scheepvaart Maatschappij N.V., all from Rotterdam.

The photograph above shows her navigating the New Waterway; below in the English waters.

*A treat for the eye: the **Vrijburgh**. She was completed by Werf De Noord N.V. in Alblasserdam and launched on 1 March 1949 and transferred to Wm H. Müller & Co. N.V. in Rotterdam on 23 May 1949. The **Vrijburgh** was deployed for the Morocco Line. She had a crew of 22 and had accommodation aboard for six passengers.*
In 1963, the open shelter deck coaster was extended by Verolme Dok- & Scheepsbouw Maatschappij in Rozenburg to increase her length from 77.34 to 85.58 metres, and her carrying capacity from 1,060 to 1,350 tonnes.

Above: Before extension in English waters.
Below: After extension on the New Waterway.

*The shipyard Firma J. Smit & Zoon from Foxhol transferred the **Follonica** to N.V. Vereenigde Nederlandsche Kustvaartbelangen (Dammers & Van der Heide), in Rotterdam, on 27 June 1949. Johannes Marten Smit from Groningen became the new owner on 6 March 1954, and renamed her **Eijo-S**. She measured a length of 44.45 metres and 394 tonnes.*

Here we see her navigating the Ems.

*Hendrik Liberg from Raamsdonkveer was the owner of the **Cornelis Houtman**. She was constructed by Firma J.G. Bröerken. Scheepswerf 'Westerbroek' from Westerbroek and completed in 1949, and named after the explorer **Cornelis de Houtman** (1565) who was the first Dutchman to sail to Java in 1595.*

The photograph shows the raised quarterdeck coaster in an English harbour.

APPINGEDAM

*These photographs show us the pretty **Appingedam**; above on a trial cruise, and below in the official promotion photograph for N.V. E.J. Smit en & Zoon's Scheepswerven, Westerbroek. The coaster was completed for Boomkamp's Scheepvaart Bedrijf in Amsterdam in July 1949. When the left photograph was taken, she belonged to Evert Smit and associates, Groningen. They bought the shelter deck vessel from Boomkamp in 1962. The **Appingedam** was 61.15 metres long and had a gross tonnage of 499 and a carrying weight of 738 tonnes.*

Following the coast

Though it looks as if her hull needs a fresh dab of paint, the ***Claes Compaen*** *is here on her maiden voyage. Here she is seen moored in Amsterdam with 130 standard shape cuts of timber from the Sundsvall District, on Monday morning 19 July 1949. The owner was Adriaan Buitelaar from Rotterdam.*

N.V. Motorschip Brabant, founded in 1927, was the owner of the ***Michel Swenden****. The coaster had a raised quarterdeck and is seen here in Swansea. She was built by Gebr. G.&H. Bodewes in Martenshoek in 1949. The ship's interests were promoted by N.V. Agence Maritime John B. Vets & Co. in Antwerp.*

*The **Henny-J**, seen here in Fowey, was built by N.V. Scheepswerf Gebr. van der Werf from Deest and transferred in December 1949 as the **Kortenaer** to Firma Hollandsche Zeerederij in Rotterdam, where the ship was in control of J. Vermaas Scheepvaart Bedrijf. The coaster had a raised quarterdeck and a carrying capacity of 530 tonnes, and was sold to G. Jonker in Groningen in January 1961 and renamed **Henny-J**, though Vermaas continued the management.*

*Wm H. Müller & Co. N.V. operated a service from Rotterdam to the Seine till Paris for a long time, and had cargos for Amsterdam and Antwerp as well. The owner also deployed ships from its English daughter company Vianda Steamship Company Ltd., London. The **Marne** was acquired for this specific purpose on 20 December 1949. When her masts were lowered she had no problem with navigating rivers with bridges.*

*The photograph shows us the **Marne** on the North Sea Canal.*

The Fifties

N.V. Reederij 'Nautiek' from Rotterdam, founded on 7 February 1938 as W. Böhmer's Scheepvaartbedrijf N.V., had beautiful ships in its fleet, and the ***Daje Böhmer*** *measuring 883 tons was a nice example. She was built by A.Vuijk & Zonen's Scheepswerven N.V. in Capelle aan den IJssel and transferred on 8 February 1950.*

The photograph above shows her in charter of Koninklijke Nederlandsche Stoomboot Maatschappij N.V., from Amsterdam which deployed the coaster often for trade with the Levant.
The photograph left shows the shelter deck vessel on trial cruise.

*The **Valbella** was christened by miss Isler at N.V. Scheepsbouwwerf v/h De Groot & Van Vliet in Slikkerveer on 6 December 1950. The shelter deck coaster had a length of 57.64 metres and a beam of 8.89 metres and belonged to N.V. Scheepvaartbedrijf 'Intercoast' (Schellen) in Rotterdam. In 1965 she was renamed the **Tridale**.*

The photograph shows the **Valbella** with grey hull. Left and under with a black hull as the **Tridale**.

*The **Dubhe** was built as the **Neeltje B** for Gebr. Broere in Dordrecht at wharf De Haan & Oerlemans' Scheepswerf in Heusden. She was completed on 31 August 1951 and was sold to P. Smeltekop & Hendrik Jan Engelsman in Delfzijl on 26 June 1963. Ten years later she was sold to Greece as the **Triton**.*

*As the **Neeltje B** in the upper photograph and as the **Dubhe** on the New Waterway in the lower photograph.*

*The **Borculo** belonged to the Amsterdam shipping company N.V. Stoomvaart Maatschappij 'Noordzee'. She was built by Scheepswerven Gebr. van Diepen N.V. in 1952. Here we see the flush deck coaster moored at her Amsterdam home port on 22 August 1964. The **Hoc Vinces** from 1943 is seen in front.*

*The **Fivel** moored in Cardiff underneath a coal-tip. Jan Bont from Groningen was the owner of this coaster with raised quarterdeck. She was built at N.V. Scheepswerf 'Foxhol' v/h Gebr. Muller in Foxhol and transferred on 8 March 1952.*
*The **Fivel** had a gross tonnage of 498 and a carrying weight of 665 tons, and a 6-cylinder 4S. C.S.A. Deutz diesel engine with a capacity of 398 bhp. The coaster was seen regularly in English harbours.*

*The **Binnenhaven** had the appearance of a yacht. She was built by N.V. Scheepswerf Waterhuizen J. Pattje in Waterhuizen and transferred to N.V. Gebr. Van Uden's Scheepvaart- en Agentuur Maatschappij, Rotterdam, on 20 March 1952. She was used for the regular weekly service for the Frabenelux Fruit Line from Dunkirk and Rouen in France to ports in French Morocco.*

*Above the **Binnenhaven** on her technical trial cruise; below on the New Waterway.*

*The **Markab-N** was built at Scheepswerf en Machinefabriek 'De Merwede' v/h Van Vliet & Co., in Hardinxveld. The trial cruise took place on 8 January 1952, after which the coaster was almost instantly converted into a refrigerated cargo vessel and finally delivered to Van Nievelt, Goudriaan & Co's Stoomvaart Maatschappij N.V. in Rotterdam on 16 April 1952. However, her performance was not satisfactory and she was sold to N.V. Maatschappij Zeevaart (Hudig & Veder) on 16 April 1958 and rechristened the **Thallo**. The cooling equipment was removed, which increased her carrying weight from 590 to 785 tonnes.*

*Disposed of her lifeboats we see the **Thallo** on the New Waterway (above) and probably in Australian waters (below).*

Equipped with a rolling Bipod mast deck the ***Muphrid-N*** *of Van Nievelt, Goudriaan & Co's Stoomvaart Maatschappij N.V. from Rotterdam arrested attention amidst coasters. She was launched on 16 September 1952, and had a length of 61.25 metres and a beam of 9.30 metres. The rolling system was no longer used after a series of accidents, and the mast was permanently secured to the foredeck.*

The photographs show the ship in English waters.

The photograph on the left shows us the **Mees Cremer** *on trial cruise on 29 November 1952. She was built with yard number 547 at Scheepswerf J.G. Bijlholt in Foxhol for J. Cremer Sr & Jr in Groningen. The flush deck vessel with a carrying weight of 470 tons was transferred to Mees Cremer Jr, also from Groningen, in 1958.*

The upper photograph was taken near Kiel Canal in Germany and we see her in the colours of Carebeka N.V., the company that managed the freighting.

Following the coast

*After the trial cruise of the **Gerry-S** on the Westerscheldt on 6 December 1952, N.V. Terneuzensche Scheepsbouw-Maatschappij in Terneuzen transferred her to N.V. Rotterdamsche Kolen Centrale in Rotterdam. She was a sturdy one for those days, with a gross tonnage of 499 and a carrying weight of 1,022 tons. With her half shelter deck she could carry much timber, mixed cargo, scrap, but also hard coal from the RKC's own mines in Wales.*

*On the left is the **Gerry-S** on the Channel and the two other photographs were taken when she navigated the North Sea Canal in about 1955.*

N.V. Scheepsbouwwerf v/h De Groot & Van Vliet in Slikkerveer completed the **Martien** *for N.V. Kustvaart-Reederij van Dudok de Wit & Co in Amsterdam on 30 April 1953.*
The shipping company, which was founded on 9 June 1941, had all its ships start with an M.

The upper photograph shows the 860-tons flush deck coaster with a load of grain berthed near the Cork Milling Company Ltd., at Victoria Quay in Cork. Below she cast anchor on the coast of south-east England.

*The **Theano**, here in Rotterdam port, and the **Trito**, here sailing the North Sea Canal, were real beauties. Both ships were built by N.V. Werf Gusto v/h Firma A.F. Smulders in Schiedam for N.V. Maatschappij Zeevaart (Hudig & Veder) in Rotterdam in 1953. These companies put the ships into service for the Holland-Ireland Line. Both ships carried mixed cargo and much fruit and cattle, and even had accommodation for 8 passengers.*

Here is the **Oosterburgh** *owned by Wm H. Müller & Co N.V. from Rotterdam navigating the New Waterway (under) and the North Sea Canal (above). The coaster with a deadweight of 760 tons was completed by Arnhemsche Scheepsbouw Maatschappij N.V. in Arnhem on 4 December 1953, and had a sturdy appearance with her so-called captain's deck and a robust Bipod mast with four 3-tons derricks.*

Jan Kuiper and Egbert Mooy from Groningen were the owners of the ***Auriga G****, a masterly example of shipbuilding by N.V. G.J. van der Werff's Scheepsbouw in Westerbroek. Her length was 48.77 metres and she was launched with yard number 268 on 12 December 1953, and transferred on 2 March 1954.*

The photograph under shows her in English waters, and left on the New Waterway.

Here we see the ***Musketier*** *almost berthed in an English harbour in about 1966. The coaster was transferred as the* ***Koh.I.Noor*** *on 30 June 1954 to Harm Salomons from Zuidlaren, and incorporated in the legal entity 'Koh.I.Noor' shipping business of Harm Salomons and partners in 1958. In May 1963, the* ***Koh.I.Noor*** *was sold to Jacobus Jan de Looff in Oosterhoogebrug, a hamlet in the province of Groningen, where she got her new name* ***Musketier****.*

The Paris trade could not be better depicted! We see the ***Realta*** *passing a bridge on the Seine in Rouen. The ship belonged to Schellen Scheepvaart en Bevrachting N.V. from Rotterdam and was launched at N.V. Martenshoekster Scheepsbouw in Martenshoek on 15 June 1954. These types of ships could reach ports far inland as they had a shallow draught and could lower their masts. The* ***Realta*** *sailed for the Franco-Goetha Line and called at ports like Paris, Rouen, Copenhagen, Esbjerg, Gothenburg, Helsingborg, Malmo and Trelleborg.*

*The **Oranjeborg**, with a carrying weight of 860 tons, was completed on 29 November 1954 and transferred to E. Wagenborg's Scheepvaart & Expeditiebedrijf N.V. in Delfzijl. She was renovated at shipyard Van der Giessen-De Noord N.V. in Alblasserdam in February 1969, where her carrying weight was increased to 1,325 tons and her original masts were replaced by two new robust masts which gave the coaster a completely different look. Where once the **Oranjeborg** looked elegant and slender, she now looked powerful.*

*The photograph above shows the **Oranjeborg** in Zaandam before conversion, and below on the Kiel Canal after conversion.*

*The **Aiglon** traded with Paris, and was the former **Gironde** of shipping company Rederij m.s. 'Gironde' from Rotterdam. She was built in 1955 at N.V. Scheepswerf Westerbroek v/h J.G. Bröerken, Westerbroek. The small flush deck coaster with accommodations for nine crew was transferred to Arend Jan Verbrugge in Portugaal towards the end of 1965, and was renamed the **Aiglon**.*

*Photograph above: As the **Gironde** on trial cruise with derricks; below as the **Aiglon**, without derricks, on the New Waterway.*

Harry Frank from Groningen was the owner of the flush deck coaster ***Aurora****, which was launched on 16 November 1955. The photograph above shows the deep-lying coaster almost berthed in Rotterdam port. The photograph below shows her with another colour (red instead of grey) on her hull in English waters. In August 1978, Harry Frank let her sail under Panamanian colours, but kept her name.*

On 22 November 1955, the ***IJsselborg*** *slid into the water at N.V. Scheepsbouw- en Reparatiebedrijf Gebr. Sander in Delfzijl for E. Wagenborg's Scheepvaart en Expeditiebedrijf N.V. in Delfzijl. The flush deck coaster was put into service on 6 February 1956 and carried much timber for Wagenborg. On 17 June 1969, Anne Visser from Groningen became the owner and changed her name into the* ***Klazina.***

Photo above: The ***IJsselborg*** *on trial cruise in freezing weather conditions.*
Photo below: As the ***Klazina*** *on the New Waterway.*

*This is the **Antilla** owned by Jannes Reint Bonninga from Groningen just after her launch at N.V. Scheepswerf Gebr. Coops in Hoogezand in 1957.*

*She had a flush deck, and was 56.28 metres long with a carrying capacity of over 800 tons. Here she was just leaving Southampton quay. Seeing behind the **Antilla** is Liberty **Remsen Heights** being taken in tow by two tugs towards sea; she was built in 1945 and measured 10,919 tons and belonged to American Export Isbrandtsen Lines Inc.*

*Right: The **Antilla** shortly after leaving Southampton.*

BERNINA

*The **Bernina** was a fine shelter deck vessel built by N.V. Scheepsbouwwerf v/h De Groot and Van Vliet in Slikkerveer for Schellen Scheepvaart en Bevrachting N.V. in Rotterdam. Her length was 56.90 metres, and with her 6-cylinder 2S. C.S.A. Sulzer diesel engine with 650 bhp she could fetch 10.5 knots.*

*The photo above show the 663 tons **Bernina** on her trial cruise on the New Waterway on 8 May 1956; the photo below shows her navigating the North Sea Canal. The **Bernina** even carried air-planes as deck cargo for the Royal Dutch Navy.*
On the left page we see the coaster riding to anchor with an airplane on deck, at the coast of Virginia Beach, Virginia, United States.

*The **Balticborg** was a fine example of shipbuilding when she slid into the water at Scheepswerf 'Welgelegen', C. Amels & Zoon in Makkum on 9 February 1957. This 2,390-tons coaster was 79.50 metres long, with a complement of no less than 23 men and mainly carried timber and salt. For her size, the **Balticborg** was considered to be a long haul coaster. However, she mainly sailed the Baltic.*

Above she is seen loading at Koninklijke Nederlandse Soda Industrie in Delfzijl, and below riding to anchor with a huge deck load of pit-poles.

*The **Imke** owned by the brothers Jan and Harm Damhof (later on, Jan was the only owner) from Delfzijl was a shelter deck coaster with a length of 67.40 metres and measuring 994 tons. She was built with yard number 743 by E.J. Smit en Zoon's Scheepswerven N.V. in Westerbroek in 1957, and intended for carrying mixed cargo and fruit.*

*Above: The **Imke** on the North Sea Canal in charter of the KNSM for sailing to countries around the Mediterranean and the Levant. Below: She was powered by a 6-cylinder 4S. C.S.A. Deutz diesel engine with 830 bhp to fetch 11.5 knots.*

*The **Beijerland** belonged to the fleet of Scheepvaart en Steenkolen Maatschappij N.V. from Rotterdam. The shelter deck coaster with a deadweight of 607 tons was completed by T. van Duijvendijk's Scheepswerf N.V. in Lekkerkerk on 18 September 1957.*
*The **Beijerland** was fitted with a cooling system and run a regular service for 't Dok in Harlingen to Goole and often carried strawboard and dairy products to this port.*
*She was sold to shipping company Rederij 'Willem Beukelsz' VI (Vroon's Handels- en Scheepvaart Maatschappij N.V.) in Breskens on 9 April 1971, renamed her **Jersey Express**, and converted her into a cattle ship.*

*Above: The **Beijerland** on the New Waterway.*

*Left: As the **Jersey Express** after being converted into a cattle ship.*

*A fine sheer and a streamlined backwards slanting bridge deck adorned the **Willi Böhmer**. The shelter deck coaster, seen here moored for being lengthened in Zaandam, was built by Cant-Navale Giuliano San Giusto in Trieste, Italy. Her trial trip was on 19 October 1957. In 1961, she was extended by 10 metres at the Dok- en Werf-Maatschappij Wilton-Fijenoord N.V. in Rotterdam, which made her total length 76.05 metres and increased her carrying capacity from 1,192 to 1,400 tons.*

*Below: The **Willi Böhmer** after extension.*

*This is the **Pluto**, seen here moored near her home base in the Haringhaven at the Westerduinweg quay in IJmuiden on 11 October 1981, owned by Cebo (Cementbouw) Marine B.V. from Heemstede (H.W.A. Brandjes) at the time. The coaster was transferred as the **Valsana** by N.V. Scheepsbouwwerf v/h De Groot en Van Vliet in Slikkerveer to N.V. Scheepvaartbedrijf 'Intercoast' (Schellen) in Rotterdam on 7 December 1957, and sold to Jacob Rusthoven and partners in Haren and rechristened **Capella** in 1958. The coaster had a length of 56.65 metres and a beam of 9.11 metres. In 1974, she was sold to H. Buitenkamp & H. Gremmer in Winschoten who named her the **Jurhil**. Not very long after, the flush deck coaster was altered by machine factory Gebr. Niestern Scheepswerven & Machinefabriek N.V. in Delfzijl and made suitable for transporting cement, and then sold again.*

*Right: The **Pluto** as the **Valsana**, which sailed for N.V. Scheepvaartbedrijf 'Intercoast' (Schellen) in Rotterdam for only a short time.*

PLUTO
HEEMSTEDE

Flush deck coaster ***Atlantis*** *was a pretty sight. She belonged to N.V. Scheepvaart Maatschappij Atlantis (Atlas & Economic Shipstores), Rotterdam, and was managed by N.V. J.Vermaas Scheepvaart Bedrijf.*

The photograph left shows the ***Atlantis*** *on her trial trip on the North Sea Canal after her completion on 11 July 1958.*
The photograph below shows the coaster for maintenance at construction firm Scheepswerf De Beer N.V. in Zaandam. The ***Atlantis*** *was 59.03 metres long and measured 905 tons. She was sold in March 1965 to Stig Gustavsson Partrederi in Skärhamn, Sweden, and kept her name.*

The photograph above shows the launch of flush deck **Wartena** by Firma G. Bijlsma & Zonen. Werf 'Volharding' in Wartena on 2 July 1958.
The photograph below shows how she left Harlingen for her trial cruise on 26 August 1958. The coaster had a length of 48.74 metres and was freighted by Firma J.J. Onnes, Cargadoors-, Scheepvaart- en Bevrachtingsbedrijf in Rotterdam. She was powered by a 5-cylinder 4S. C.S.A. Industrie diesel engine with a bhp of 375 to fetch 9.5 knots. The **Wartena** was named after her shipping company, and when she was sold to Sweden in April 1967 she kept her name. In 1991, after a reconstruction, the **Wartena** served as a training ship for the nautical college in Karlskrana in Sweden.

*The **Nusakan** of Van Nievelt, Goudriaan & Co's Stoomvaart Maatschappij N.V. in Rotterdam was a so-called long-haul coaster. The **Nusakan** was the first ship that was built for Van Nievelt with a raised foredeck; the second one was the **Sheratan**.*
*The **Nusakan** was completed at shipbuilding yard Scheepswerven Gebr. van Diepen N.V. in Waterhuizen on 25 September 1958, and had a length of 71.50 metres and a carrying capacity of 1,329 tons. The ship and her complement of 19 hands were running regular services for Van Nievelt on European waters and around the Mediterranean.*
She was transferred to the German subsidiary Van Nievelt, Goudriaan & Co's Stoomvaart Maatschappij (Hamburg) GmbH, Hamburg, and was not renamed.

*The photograph above shows the **Nusakan** on the North Sea Canal; below in English waters.*

Another full-bodied coaster was the ***Argonaut*** *belonging to N.V. Amsterdamse Scheepsexploitatie Maatschappij 'Argo' from Amsterdam, which was founded on 9 November 1950. Comm. Venn. Terwogt & Lagers acted as the carrier.*
The ***Argonaut*** *was 66.22 metres long and was launched by Bodewes Scheepswerven N.V. v/h Gebr. G.&H. Bodewes in Martenshoek on 14 August 1958 and her trial cruise took place on 7 October 1958.*
The ***Argonaut*** *had a raised quarterdeck and a carrying capacity of 1,393 tons. The coaster was sold under distress to N.V. Exploitatie Maatschappij 'Mavi' in Amsterdam on 20 November 1967 and not renamed. In May 1968, she was resold to Compañia Naviera Baan S.A. in Monrovia and rechristened the* ***Erasmus****. Her manager was J. Vermaas Scheepvaart Bedrijf N.V. in Rotterdam, who bought the ship in 1970, renamed her the* ***Houtman*** *and incorporated her in the company J. Vermaas Reederij Bedrijf N.V.*

Above: The ***Argonaut*** *on the North Sea Canal.*
Middle: Making the English harbour.
Right: As the ***Erasmus*** *arriving in Leixoes.*

The **Niña**, owned by Dammers & Van der Heide's Scheepvaart- en Handelsbedrijf N.V. from Rotterdam was the namesake of one of the ships Columbus used on his voyages. She was launched as a half shelter-deck coaster at Bodewes Scheepswerven N.V. v/h Gebr. G.&H. Bodewes in Martenshoek on 11 March 1959, and transferred to N.V. Motorscheepvaart Maatschappij 'Rotterdam' (Dammers) in June of the same year. The coaster was converted to have a full shelter deck. In 1965, the **Niña** was transferred to N.V. Motorscheepvaart Maatschappij 'Columbus' in Willemstad on the Netherlands Antilles, but Dammers continued the management.

The photograph above shows the **Niña** on trial cruise; the photograph below after she was converted into a coaster with full shelter deck on the New Waterway.

The sixties

The reefer ***Geestland*** *was built at N.V. Scheepswerf 'Welgelegen', Firma C. Amels & Zoon in Makkum. In November 1959 the ship was christened by Mrs. Lucille van Geest, wife of John van Geest, president of Waling van Geest & Zn. N.V., the ship's owner. And in April 1960 she set off for her maiden voyage as a member of the Geest Line, to the Windward Islands to load bananas. In those days, it often were the women who brought the bananas on board, walking over the gangway and carrying the bunches of bananas on their heads. The fine reefer was able to transport 100,000 stocks of bananas. In 1971 the* ***Geestland****, which measured 1,937 tons, was sold to to Harvard Shipping Co. Inc., in Panama and she was renamed* ***Banana Planter****.*

Above: Launch of the ***Geestland*** *at N.V. Scheepswerf 'Welgelegen' in Makkum.*
Below: On her way to the Windward Islands.

The half-shelter deck ship ***Agua*** *was launched at N.V. Nieuwe Noord-Nederlandsche Scheepswerven in Groningen on 2 June 1960. The trials and delivery to shipowner Jannes Reint Bonninga followed on 14 July 1960.*
At the time, this shipowner also had the ***Aruba*** *(1951) and* ***Antilla*** *(1957) sailing in his service. The 64.52 metres long and 10.32 metres wide coaster appeared to be longer than she actually was, due to the short well deck and the nicely streamlined superstructure thanks to the funnel, the cap of which continued over the wheel house.*
The ***Agua*** *had a gross tonnage of 499 and a deadweight capacity of 1,157 tons. Attached to the aft mast was a derrick that could lift between 3 and 15 tons, attached to the fore mast were two derricks of 3 tons each.*
The ***Agua*** *was managed by M. Havinga's Scheepvaart- & Handelmaatschappij N.V. in Groningen. In 1974 the* ***Agua*** *was sold to Rederij 'Watum' of Jan Hzn. Damhof in Delfzijl and she was renamed* ***Watum****.*

Above: The ***Agua*** *photographed on the Ems in northwestern Germany even before the official trials on the 14th of July 1960.*
Middle: Sailing in English waters.
Below: As ***Watum*** *of Jan Hzn. Damhof.*

On 15 February 1960 **Sint Annaland**'s keel was laid at Bodewes Scheepswerven N.V. in Martenshoek. The coaster of the open shelter deck type was launched on the 30th of June 1960 and on the 20th of October of that same year she was delivered to her owner N.V. Scheepvaart- en Steenkolen Maatschappij in Rotterdam. The coaster had a length of 68.46 metres and a breadth of 10.04 metres, two decks and two cargo holds of which 301 m^3 was refrigerated/freezing hold. The gross tonnage amounted to 499 and the deadweight capacity to 828 tons. The ship had a 8-cylinder 4 S. C.S.A. Werkspoor diesel engine of 1,100 bhp. The amount of crew on board could vary between 15 and 18. The **Sint Annaland** was an often welcomed guest in the United Kingdom. However, because the S.S.M. faced severe competition of other shipowners who employed container and ro/ro ships on the routes to the United Kingdom, the conventional **Sint Annaland** had to be put in an other service. S.S.M. found the solution in establishing a subsidiary in South Africa, where services along the South-African coast created opportunities. In February 1968, **Sint Annaland**'s flag became South-African when she sailed for the Cape Natal Line (Pty) Ltd., (S.S.M.) in Capetown.

Above: During trials of the **Sint Annaland**.
Below: Assisted by the small Wagenborg tug **Golfbreker**, the **Sint Annaland** enters Delfzijl after her trials.

Rederij Anne Herfurth N.V. of Rotterdam had ordered Bodewes Scheepswerf Foxhol N.V. in Foxhol to build the flush deck ship **Anne Herfurth**. The coaster ran down the slips athwartships on 8 October 1960. The modest ship had a length of 58.35 metres between perpendiculars and of 61.08 metres length overall. Her breadth was 9.25 metres and the depth 3.08 metres. The management of the coaster, measuring 965 tons, was dealt with by the company Herfurth Scheepvaart- & Transportbedrijf N.V. in Rotterdam, which was established in 1950 and which itself was part of the firm Herfurth & Co. that was originally established in Antwerp in 1899.

*Above : The **Anne Herfurth** under construction at the shipyard.*
Left: A beautiful photo, probably taken for the official trial trip.
Right: Moored in an English harbour.

The next page:

*In the fifties and sixties of the 20th century, E. Wagenborg's Scheepvaart- en Expeditiebedrijf N.V. of Delfzijl had numerous, almost identical coasters in service. Two of those were the **Dintelborg** and **Dongeborg**. These flush deck ships were mainly built for the transport of timber products from the Scandinavian countries and Sweden and Finland in particular. Naturally, they would be chartered for other cargo markets as well when demanded, but essentially it was North-West Europe.*

*The **Dintelborg**, pictured above, was built at N.V. Scheepswerf 'Friesland' in Lemmer where she was launched on 10 March 1961 and delivered on the 25th of April 1961. The coaster's length was 46.01 metres and her deadweight capacity was 513 tons.*
*The **Dongeborg**, pictured below before the official sea trials on the 4th of October 1961, was a product of N.V. Scheepswerf 'Appingedam' v/h A. Apol in Appingedam and she was launched athwartships on 24 August 1961. Both ships had a 4-cylinder 4S. C.S.A. Brons diesel engine of the type 4ED of 220 bhp, which enabled the ships to sail with a speed of eight to nine knots.*

*On 24 June 1975, the **Dintelborg** was sold to Geert and Willem Timmer in Groningen, who renamed her **St Michael**. The **Dongeborg** had already been sold earlier, on the 8th of April, to Pieter Pot in Delfzijl, Hindrik Pot in Kampen and Pieter Pot in Siddeburen, each having a share of 1/3, and they renamed her **Vissersbank**.*

DINTELBORG

DONGEBORG

U
COCCINELLE
COCCINELLE
HENSEN
ROTTERDAM

The photo on the left shows us the **Coccinelle**, *French for ladybird, just before the launching at IJsselwerf N.V. in Rotterdam on the 22nd of February 1961. The completion of the open shelter deck ship, which originally would get the name of* **Waalhaven**, *was carried out at Werf De Hoog N.V. in Rotterdam. After trials on 25 May the coaster, with a length of 67.32 metres and a breadth of 10.39 metres, was delivered to N.V. Scheepvaart Maatschappij 'Waalhaven', a shipowner that was part of Gebr. Van Uden's Scheepvaart- en Agentuur Maatschappij in Rotterdam, on the 1st of June 1961.*
Sistership of the **Coccinelle** *was the* **Yselhaven**, *delivered on 8 May 1962. Van Uden put both ships in service in the Frabenelux Ligne. This was a weekly scheduled service from Dunkirk and Rouen in northern France to the West-Moroccan ports of Casablanca, Kenitra and Safi.*

The photo above shows the trials and the one below shows the ship sailing on the New Waterway.

An eyecatcher in the Dutch coastal shipping fleet was the ***Bellatrix*** *of the shipowner Bernard van der Laan of Schiedam. The ship was equipped with two goal post masts and also her aft superstructure differed slightly from the average ones as seen in the Dutch coastal shipping fleet in those days. On 9 May 1961 the* ***Bellatrix****, a ship of the shelter deck type, was launched at N.V. Scheepswerf 'Vooruitgang' Gebr. Suurmeijer in Foxhol as* ***Marijke Irene****. However, on her trail trip on the Ems in July the name of* ***Bellatrix*** *was already stated on her bow. Straight after the delivery the coaster was taken in bare-boat charter for ten years (hence the change of name) by Van Nievelt Goudriaan & Co's Stoomvaart Maatschappij of Rotterdam. After her return to her original shipowner on 26 April 1971, she was rebuilt for the transport of heavy cargoes and her flag became that of the Netherlands Antilles.*

*The chemicals tanker **Stella Polaris** was built as yard number 315 at N.V. Noord-Nederlandsche Scheepswerven in Groningen.*
On the 10th of May 1961 the tanker was launched and christened by the wife of the managing director of N.V. Reederij 'Theodora' in Uithoorn, Mrs. C. van der Post-Vink.
The 60.66 metres long and 8.95 metres wide tanker was delivered to the shipowners on 18 July 1961 and she made numerous voyages in charter in the chemicals service between ports in northwestern Europe and the Mediterranean.
*Just after 10 years the 910 tons **Stella Polaris** was sold already and in 1971 she transferred to Navigazione Adriatica SrL in Rimini, Italy and she was renamed **Giaro D**.*

Above: Trials on 18 July 1961.
Middle: Sailing on the New Waterway for the first time.
*Below: As **Giaro D** in La Valletta on Malta.*

On 24 July 1961 the reefer ***Echo*** *of N.V. Maatschappij Vriesvaart in Rotterdam, established in 1959, was delivered. This shipowner was part of the firm Hudig & Veder N.V. which had already been established as far back as 1795. The* ***Echo*** *was built at N.V. Scheepwerf 'De Hoop' in Lobith and on the 16th of March 1961 she was committed to the seas. The trial was carried out on 17 July 1961. The beautiful white ship had three refrigerated cargo holds, each with a capacity of 2,101 m*^{3}*.*
The gross tonnage of the ship, which had a length of 88.40 metres, was 987, due to changes in the regulations this was altered into 1,830 tons later on. The deadweight capacity first amounted to 1,750 tons and later on this increased to 2,170 tons.
The ***Echo*** *had a crew of 26 and already after nine years she was sold to H/f Eimskipafelag Islands in Reykjavik and renamed* ***Ljosafoss****.*

Left page: The aft ship of the ***Echo****.*
Above: In the English Channel.
Below: Sailing on the New Waterway.

NASSAUBORG

NASSAUBORG
DELFZIJL

*On 7 September 1961, the **Nassauborg** of E. Wagenborg's Scheepvaart & Expeditiebedrijf N.V. from Delfzijl was launched at N.V. Scheepswerf 'Welgelegen', C. Amels & Zoon in Makkum and on the 14th of December 1961 she was delivered to the shipowners. De shelter deck ship had a gross tonnage of 499 and a deadweight capacity of 985 tons. In January 1969 the coaster was rebuilt and lengthened at Van der Giessen-de Noord N.V. in Alblasserdam, which increased the gross tonnage to 1,384 and the deadweight capacity to 2,245 tons.*

After the lenghtening, her overall length became 81.86 metres, and that was 69.76 metres originally. Between perpendiculars the length remained 74.43 metres. All this had to do with the introduction of the 75 metres limit. Due to a change in the law, instead of her tonnage it would be the length of a ship that would now determine whether a ship could be categorized as fit for ocean-going service or for short-haul service.

*The **Nassauborg** was sold to Pondo Chartering Inc., in Panama and renamed **Attiki** on 16 April 1980.*

*Above left: The **Nassauborg** moored in Zaandam. The tarpaulins have been raised, which indicates that the discharging of the timber was stopped temporarily because of a rain shower.*
Bottom left: Leaving Delfzijl on her way to Norway.
Above and below, after her reconstruction she is seen on the New Waterway.

TEXELSTROOM

TEXELSTROOM

*The **Texelstroom**, owned by N.V. Hollandsche Stoomboot Maatschappij in Amsterdam, was a ship that was built according to a special plan in order to be able to load and to discharge more efficiently. Built at Arnhemsche Scheepsbouw Maatschappij N.V. in Arnhem, the **Texelstroom** was launched on the 24th of May 1962. The trials of the 79.22 metres long and 11.43 metres wide coaster took place on 10 August 1962 and the day after that she was delivered. The **Texelstroom** was a vessel of the open shelter deck type and she measured 1,348 tons. Instead of derricks she had two deck cranes that were able to lift between three and six tons each. The ship was put in the Hull service. In 1968 she was transferred to the Mersey service, Amsterdam/Rotterdam/Antwerp/Liverpool v.v. At a later date the ship, then sailing as part of the N.V. Nederlandsche Scheepvaart Unie in Rotterdam (Holland Bulk Transport N.V.), would also sail in charter for a year for the Compagnie Nationale Algériënne de Navigation in Algiers. She also sailed in charter for Ellerman Lines Ltd., in Londen for a short while.*

*Above: The **Texelstroom** on the North Sea Canal.*
Above left: Along the quay in Plymouth on 9 April 1973.
Bottom left: Entering Dover, 1973.
*Below: After her sale in May 1976, the coaster successively sailed with the names **Athanassios-S**, **Julia**, **Lucky C** and finally **Florida** on her bow, for various shipowners. The photo shows her as **Florida** at the shipbreakers in Aliga in December 1995.*

Hendrik Damhof of Delfzijl was the shipowner of the ***Hoop****. This coaster of the shelter deck type was built with yard number 109 at Scheepswerf Hoogezand N.V., Jac Bodewes in Hoogezand. The* ***Hoop****, registered for Ice Class 3 was launched on 5 July 1962 and on the 5th of October that year she was delivered. The engine room housed a 8-cylinder 4S. C.S.A. Klöckner-Humboldt-Deutz diesel engine of 1,000 bhp, sufficient for a speed of 12 knots. Shipbroker of the 62.41 metres long and 9.81 metres wide coaster was N.V. Carebeka in Groningen. In 1971, B.V. Scheepvaartbedrijf Poseidon from Delfzijl became her new shipbroker.*

*The **North Pole** and the **South Pole**, two identical sisterships of 1,589 tons each were built as **Coral Acropora** and **Coral Actinia** respectively at Chantiers & Ateliers Aug. Normand in Le Havre for Koraal Scheepvaart Maatschappij N.V. (Anthony Veder) in Willemstad, Netherlands Antilles.*
*The launch of the **Coral Acropora** took place on the 1st of August 1962 and in October 1962 her delivery followed.*
*The **Coral Actinia** was delivered on 11 May 1963. The shelter deck type reefers sailed in bare-boat charter for Dammers & Van der Heide's Scheepvaart en Handelsbedrijf B.V. in Rotterdam and their voyages predominantly took them to the United States of America. In 1973 the ships were sold to Reefer Express Co. N.V. in Willemstad (Vroon) and they were renamed **Frio Express** and **Iglo Express** respectively. In 1977 the **Frio Express** was renamed **North Pole** with Rederij 'North Pole' N.V. in Willemstad as her owner, while the **Iglo Express** became the **South Pole**, now owned by Rederij 'South Pole' N.V. in Willemstad. These two shipowners too belonged to Vroon's Handels- en Scheepvaart Maatschappij N.V. in Breskens.*

*Left page: On the photo below we see the **Hoop** in English waters and above we see her entering Hamburg.*
*Above right: The **South Pole** on the New Waterway.*
*Right: The **North Pole** moored along the Noorderkade in the port De Pijp in Beverwijk on 19 February 1985.*

On 22 May 1963, the **Kittiwake** of P.A. Van Es & Co N.V. of Rotterdam was delivered. She was built as yard number 1304 at N.V. Scheepsbouwwerf 'De Vooruitgang' v/h D. Boot in Alphen aan de Rijn. The coaster, launched and christened by Mrs. Th. van Zelm-van Eldik as **Breewijd** on the 7th February 1963, was of the open shelter deck type. Before the delivery her name had been changed in **Kittiwake**. Her length was 72.67 metres and her breadth 10.42 metres, with a gross tonnage of 496 and a deadweight capacity of 1,219 tons. The ship had a 6-cylinder 4S. C.S.A. Industrie diesel engine of 1,350 bhp, enough for 13,5 knots of speed.

The **Kittiwake** had two decks, one cargo hold, two hatches and six derricks of 3 tons each that were attached to two goal post masts. Accommodation was available for a crew of 12-15. On 5 December 1974 she was sold to Partrederiet Svenborg Grace and her management transferred to Svendborg Enterprise ApS in Svendborg, Denmark. She was renamed **Svendborg Grace**.

Above: The **Kittiwake** on 21 May 1963, seen on the New Waterway on her way to the North Sea for her trial trip.
Below: As **Svendborg Grace** moored along the Noorderkade in the port De Pijp in Beverwijk to load seed potatoes, on 13 October 1977.

The picture above shows the trials of the ***Sonja****, still in her original shape. Later on the boat deck was slightly reconstructed and a second life boat was installed. The shelter deck ship, launched on the 29th of March 1963, was built at N.V. Nieuwe Noord-Nederlandsche Scheepswerven in Groningen as ordered by Rederij m.s. 'Sonja', Irish & Continental Shipping Co. Ltd., in Dublin. This was the company of the shipowner H. Pinkster from Groningen, and Groningen also became the homeport of this twin screw coaster. The 78.36 metres long reefer was bought by Van Nievelt, Goudriaan & Co's Stoomvaart Maatschappij N.V. of Rotterdam on 7 March 1973, who renamed her* ***Meres****. The lovely coaster with her bridge amidships, only sailed for a short period of time for the Rotterdam office of this shipowner. On the 13th of November 1976 she was transferred to the Greek daughter company Dionikos Shipping Co. Ltd., in Piraeus and renamed* ***Frigo Queen****.*

Photo below: As ***Meres*** *on the New Waterway.*

Following the coast

A beauty of a coaster was the ***Twee Gezusters*** *owned by Jan sr and Harm Damhof. This shelter deck ship was launched at E.J. Smit en Zoon's Scheepswerven N.V. in Westerbroek on 5 April 1963. The delivery of the 74.30 metres long and 10.44 metres wide coaster followed on the 18th of June. She was a real liner coaster and amongst others she was used a lot for the transport of fruit from countries along the Mediterranean to northwestern Europe. On board a crew of 11 to 16 would fit easily. The gross tonnage amounted to 499 and the deadweight capacity to 1,062 tons. Six derricks were attached to two masts, two of 3 tons and four of 5 tons.*
After Harm Damhof sr had passed away in 1966, the Damhof heirs became the owners of the ship, the widow Damhof-Schrik and Harm Hzn Damhof.

Left page: The engine room of the ***Twee Gezusters****, on the photo leaving the Zandvlietsluis (lock) in Antwerpen, housed a 16-cylinder 2S. C.S.A. Appingerdammer Brons diesel engine of 1,250 bhp.*
Above right: Entering Delfzijl after the trial trip.
Middle: On 26 September 1977 the ***Twee Gezusters*** *moored along the Noorderkade in the harbour of Beverwijk. The ship was loading seed-potatoes with a conveyer belt, as was common practice in those days.*
Below: The ***Twee Gezusters*** *on the New Waterway.*

Pictured above is the **Carebeka III** before she was lengthened, upstream on the Gironde just before mooring in the small town of Libourne near Bordeaux.
The flush deck ship was launched at N.V. Scheepswerf Waterhuizen J. Pattje in Waterhuizen on 3 July 1964 and she was delivered on the 20th of November 1964.
The 63.86 metres long and 9.58 metres wide coaster had a gross tonnage of 499 and a deadweight capacity of 982 tons.
In October and November 1969 the **Carebeka III** was rebuilt and lengthened at Niestern's Scheepswerven Groningen N.V. in Groningen, resulting in an increase in length to 71.02 metres and in deadweight capacity to 1,772 tons.

The photo right shows the lenghtened **Carebeka III** in estuary of the New Waterway mid 70s.

N.V. Carebeka (Capitein-Reders-Bevrachtings-Kantoor) (Captain-Shipowners-Shipbrokers-Office in English) was established in Rotterdam on 23 May 1939. The founders were a number of captains-owners working in the coastal shipping, who decided to join forces in a company managed by themselves as well as taking care of their own cargoes by being their own shipbroker. Their intention with this initiative was to end the wrongs in the world of shipbrokers and also to lower and control the costs they continuously faced. An other reason of course was that when everything could be managed by themselves the competition between them would disappear. In short, as shipbroker with a large number of captains-owners they were a much more solid power against the existing world of brokers and what's more, the profits made would remain within their own company too. Their office was settled at the address Emmaplein 1 in Groningen. At first only two rooms were rented but after 1946 they rented the whole premises and they would buy it all in 1954.

Over the years, many shipowners would join this firm, which became a household name in the world of coastal shipping. Their business flourished such that it was decided to open a branch office in Rotterdam, on 10 August 1953, followed by a further branch in Delfzijl on 15 December 1955.

In 1958 the Carebeka organisation itself became a shipowner, when they put their own ship in service. This was the **Carebeka***, how originally thought, and she was put in service for Rederij 'Carebeka' in Groningen on the 5th of February 1957. Nine ships would follow, all with the prefix Carebeka followed by a Roman numeral. As from 1959 they even introduced their own logo on the funnel: the various ships within the joint venture would now sail with a yellow funnel with two green bands around it with a white band in between. In the white band a logo was visible of a green circle with a white triangle and a black letter C therein. The company's own ships had a white funnel instead of a yellow one, while the remaining colours remained the same.*

In July 1972 the Rotterdam office became completely independent and it would continue its business as Carebeka Transport (Rotterdam) B.V. In this office Carebeka participated for 50% while the other 50% was registered to Granaria Rotterdam. This office too decided to sail services using their own ships and the first of those was the **Hunnau***, bought in Germany she was put in service as* **Francinaplein** *in July 1973. Four ships in total would be put in service by this office. The funnel logo was the same as that of Carebeka Groningen except for the fact that the bands and logo were now red instead of green.*

Unfortunately, Carebeka's business went downhill. It was no longer a joint venture which represented the interests of the captains-owners in it. Due to the fact that Carebeka now also had their own ships built, it had to borrow capital on the financial markets to finance this and they faced more and more difficulties to pay the accompanying interest. This would finish off the shipowners and they were declared bankrupt on 28 January 1983.

In November 1974, Carebeka Transport (Rotterdam) B.V. bought the ***Marmora*** *of the Dutch shipowner W. Koerts & Zoon in Zuidbroek. She had been built as* ***Marmorfjell*** *for I/S Marmorfjell (Hvide & Schjött) in Bergen, Norway at Kristiansands M/V A/S, in Kristiansand where she was delivered on 4 March 1965. The 71.73 metres long and 11.51 metres wide coaster was given the name* ***Calandplein****. The ship, measuring 2,230 tons, had a 8-cylinder 4S. C.S.A. Atlas-Mak diesel engine of 1,400 bhp, sufficient for a speed of 12.5 knots. Carebeka sailed her in tramp service.*
In October 1979 Carebeka already sold the ship, to Dernal Shipping Co. Ltd., in Limassol, Cyprus (Nicolaos Hadjidakis, Pireaus) which renamed her ***Anthoula I****.*

The photo above shows the ***Calandplein*** *on the New Waterway and on the one below she is seen as* ***Anthoula I*** *in La Valletta, Malta.*

The RoRo ***Rynstroom*** *of N.V. Hollandsche Stoomboot Maatschappij from Amsterdam was the first roll-on-roll-off ship for this shipowner. Built at Arnhemsche Scheepsbouw Maatschappij in Arnhem she was launched on the 30th of October 1965, followed on the 1st of February 1966 by her trial trip on the IJ waters in Amsterdam and her delivery the next day. The ship had a length of 74.52 metres and she was of the closed shelter deck type. On 5 February 1966 she set off for her maiden voyage in the service Amsterdam-Shoreham v.v. Later on, voyages to the Mediterranean would follow too. In March 1976 she was sold to Handels & Scheepvaart Onderneming Vroon B.V. in Breskens and renamed* ***Car Express****. In 1976 the ship was lengthened with 6.64 metres and in 1977 she was rebuilt to become a transport ship for live cattle.*

Above: The ***Rynstroom*** *returns from her trial trip on the IJ waters in Amsterdam on 1 February 1966.*
Middle: Sailing in the English Channel after her reconstruction in 1967; during which amongst others her aft mast had been removed, the top aft deck continued and a loading ramp aft had been installed.
Below: As ***Car Express*** *sailing for Vroon in Breskens. Her hull now was grey instead of black.*

*Hendrik Damhof Jr uit Delfzijl was the owner of the fine flush deck ship **Jan**. This coaster, the successor of the **Jan** built in 1952, was yard number 133 of Scheepswerf Hoogezand N.V., Jac. Bodewes in Bergum where she was launched on 16 November 1966. The coaster had a length of 73.05 metres and a breadth of 10.32 metres, and she had a crew that could vary between 9 and 14. Her gross tonnage was 512 and the dead-weight capacity 1,063 tons. Although the **Jan** had a reinforced bow (Ice Class 3) in order to be able to sail through ice fields, she mainly sailed to countries around the Mediterranean, the Levant and North Africa transporting fruit. The coaster, equipped with a tweendeck, was also used for the transport of seed potatoes from the Netherlands to the countries mentioned above, returning to north-western Europe with cargoes of edible potatoes.*

*Above: The **Jan** on the New Waterway.*
Below: Sailing on the Mediterranean, fully laden 'up to her ears'.
*Right page above: The **Jan** seen at night while moored along the Noorderkade of the port De Pijp in Beverwijk loading seed potatoes on 13 January 1978.*
Page right below: The coaster sailing on the North Sea Canal near Amsterdam.

JAN

JAN

The ***Schildmeer****, also owned by Hendrik Damhof Jr of Delfzijl, was the former* ***Looiersgracht*** *of Spliethoff's Bevrachtingskantoor N.V. in Amsterdam.*
In 1975 he bought the shelter deck coaster, built at N.V. v/h Scheepswerven Gebr. van Diepen in Waterhuizen where she had been launched on 24 March 1966 followed by her trials on 28 April 1966.
The ***Schildmeer****, with a length of 72.94 metres and a breadth of 10.83 metres had accommodation for a crew of 9-11. The coaster, equipped with a tweendeck, had a gross tonnage of 490 and a deadweight capacity of 1,168 tons. Later on, due to changing regulations, this was changed into a gross tonnage of 1,111 and a deadweight capacity of 1,876 tons.*
The ***Schildmeer****'s shipbroker was N.V. Scheepvaart Bedrijf Poseidon in Delfzijl.*

Both photos were taken in the port De Pijp in Beverwijk, where we see the ship moored along the Noorderkade loading seed potatoes.
Above: 4 November 1982.
Below: 3 October 1977.

The ***Seinehaven*** *(1967) was bought by Gebr. Van Uden's Scheepvaart- en Agentuur Maatschappij in Rotterdam from Société Navale Caennaise, Caen as* ***Niobe*** *on 11 October 1973. The ship, 78.19 metres long and measuring 1,750 tons was able to transport refrigerated cargo and amongst others she was put in service in the Frabenelux Ligne. She had been launched at N.V. Scheepswerf 'De Waal' in Zaltbommel on the 10th of March 1967.*
The reefer only sailed for Van Uden for three years; in October 1976 she was sold to A. Martinez y Compania in Santo Domingo and she was renamed ***Rio Yaque****.*

The photos show the coaster on the New Waterway.

Based on positive results, Carebeka B.V. of Groningen decided to purchase four ships from Norway. One of these was the ***Lionel*** *of I/S Protector (Alf Mohn, managers) in Oslo, bought on 31 January 1971. This robust coaster had been built at Stocznia Gdanska im Lenina in Gdansk, which had delivered her on 29 December 1967. Rederij 'Carebeka' renamed her* ***Carebeka VI****.*
The 74.27 metres long and 13.06 wide coaster had a gross tonnage of 1,289 and a deadweight capacity of 2,280 tons. The coaster had two derricks of ten tons each attached to a goal post mast in the middle of the ship, and two of 5 tons each which were attached to a goal post mast in front of the bridge and a mast on the forecastle.
On 30 December 1982 the ***Carebeka VI*** *was sold to Linde Loyd B.V. from Ursum and renamed Berkborg (in bare-boat charter for Wagenborg), after that she would change ownership a couple of times more.*

Above: Empty, just after leaving the Middensluis (lock) in IJmuiden on her way to Amsterdam.
Middle: The ***Carebeka VI*** *still with her dark hull, passes the Koningsbrug (King's Bridge) in Rotterdam.*
Below: As ***Berkborg*** *in Rotterdam on 14 July 1983.*

Moored along the Parkkade (quay) in Rotterdam with a glimpse of the former Holland-America Line building at right in the background.

*The laying of the keel of the chemicals tanker **Dutch Faith** of N.V. Tankvaart Rotterdam in Rotterdam - as from 1972 named N.V. Tankvaart Dordrecht in Dordrecht (Gebr. Broere N.V) - took place on the 19th of June 1968 at N.V. Scheepsbouwwerf v/h De Groot & Van Vliet in Slikkerveer.*
*Already during construction the tanker was lengthened; this was caused by the new regulation of the maximum of 75 metres length between perpendiculars. The launch of the **Dutch Faith** took place on 30 November 1968 and the delivery to her owner followed on 3 March 1969.*
*The **Dutch Faith** had a length of 69.43 metres and a breadth of 11.53 metres. The gross tonnage amounted to 999 and the deadweight capacity to 1,880 tons. The engine room housed a 8-cylinder 2S. C.S.A. Smit-Bolnes diesel engine of 1,760 bhp. The tanker mainly sailed in European waters and frequently she called at the English port of Teesport, located at the southern bank of the river Tees.*
*In January 1996 the **Dutch Faith** was sold to Bergen Tankers A/S (Olav Hauglund) in Bergen, Norway and renamed **Bergen Faith**.*

Above: Following her sale to Broere Bros. her funnel became black with a blue band with the letters GB in white in it.
*Below: The **Dutch Faith** on the Westerscheldt on her way to Antwerp on 10 March 1977.*

The seventies

*On 15 June 1970 the keel of the **Vincent** was laid at Bodewes Scheepswerven N.V. in Martenshoek. On 20 November 1970 the launching took place and the delivery to shipowners Rederij 'Vincent' in Groningen was on 8 April 1971. She was managed by Beck's Scheepvaart-kantoor N.V. in Groningen.*
The coaster with a raised quarterdeck was 55.10 metres long and 9.25 metres wide. Her gross tonnage was 399 and the deadweight capacity amounted to 766 tons.
*The **Vincent**, which regularly sailed to England, had a 6-cylinder 2S. C.S.A. Appingedammer Brons diesel engine with 450 bhp, enough for 11 knots. The coaster had one cargo hold and two derricks, one of three tons and one that could lift between three and five tons.*

*On 15 September 1986 the **Vincent** was sold to Albert Pahladsingh in Voorschoten, who sold her again in Januari 1987. At the moment she still sails with the same name and under Dominican flag for Vincent Shipping Co., Kingstown, St Vincent & the Grenadines and she is managed by Suriname Coast Traders in Paramaribo.*

Above: The raised half deck can be well recognised.
*Below: The **Vincent** at the Parkkade (quay) in Rotterdam on 14 July 1983.*

The **Atlantic River** of the Oost Atlantic Lijn B.V. in Rotterdam was built as **Norimo** at Aukra Bruk A/S in Aukra in Norway. The keel was laid in September 1970 and in July 1971 the ship was delivered to Normann Shipping A/S (Rolf Normann) in Oslo. That same year A/S Norexim (Norexim Shipping A/S) in Oslo became the new owner of the **Norimo**. In May 1974 the 77.71 metres long and 13.03 metres wide ship was sold to shipowner 'Munte' in Haren, of the gentlemen J. Bont and H.R. Dost and renamed **Munte**. In March 1980 an other sale of the **Munte**, measuring 2,489 tons, followed to Sylvia Cargo B.V. in Groningen and now she was managed by bij Otto A. Muller in Hamburg. Next the Oost Atlantic Lijn became the new owner on 18 May 1981 and the coaster was renamed **Atlantic River** with Rotterdam as her homeport. On 28 August 1986 her flag became a flag of convenience and Limassol became her new homeport. She now sailed for Adamant Marine Ltd., in Nicosia on Cyprus. The management remained at the Oost Atlantic Lijn. The **Atlantic River** had one cargo hold, two hatches and four derricks of 5 tons each. In December 2001, following engine damage due to a broken connecting rod off the south coast of England, the ship was towed to the Madroelhaven in Pernis by the Firm Broedertrouw. On 18 January 2002 the ship was officially withdrawn from the registers at the O.A.L. Early February 2002 the coaster was sold to Dick van der Kamp Shipsales B.V. in Brielle and towed to the former shipyard Van der Giessen-De Noord N.V. in Alblasserdam. Here the engine was repaired and a new crankshaft was installed as well as a new top motor block (MaK 8MU452AK). In March 2003 she was sold to Liverpool Navigation Pvt, Ltd., in Colombo on Sri Lanka and renamed **City of Dublin**.

Right: A view from the bridge wing on star board towards the fore ship, looking at the loading of the seed-potatoes in the port De Pijp of Beverwijk on 8 December 2000.
Above: The ship as seen from the quay during loading on the same date.
Below: With homeport Limassol on her rear in the Middensluis (lock) in IJmuiden on 2 August 2001.

BEVEROL
CAPE COD

BREEHEES

BREEHEES

*The **Breehorn** and **Breehees** are almost identical sisterships that were built at Batservice Verft A/S in Arendal and Mandal in Norway respectively.*

*After trials in October 1971, the **Breehorn** was delivered to Lys Line A/S (Rederi Lystrade) as **Lyshav** and she was then managed by Simonsen & Slang A/S in Oslo. The ship was 80.02 metres long and 13.76 metres wide. Her gross tonnage amounted to 1,585 and the deadweight capacity to 3,036 tons. In 1975 she was sold to P.A. Van Es & Co in Rotterdam and renamed **Breehorn**. In July 1984 followed the sale to Porthos Shipping Co. Ltd., in Piraeus and the coaster was renamed **Porthos**.*
*In 1985 Seabaron Shipping Co. Ltd., in Limassol became the new owners and the name **Chrysanthi S** appeared on the bow. In 1992 she was sold to Prudentmar Ltd., in La Valletta on Malta and she was renamed **Chrissi P**. In 1997 she became the **Fos** of Sea Dream Naviagation Ltd., in La Valletta. The last registration dates from 17 February 2003 in which it is stated that she is a so-called dead ship. Her name now was **Rose S** (Honduran flag) of Libra Shipping in Athens and managed by Sefy Shipping & Trading in Athens.*

*The **Breehees** was launched on 4 February 1972 and delivered to A/S Bleikvassli Gruber in Moirana, Norway as **Vaddas** in June 1972. The coaster was 79.96 metres long and 13.76 metres wide. Her gross tonnage amounted to 1,586 and the deadweight capacity to 2,997 tons. P.A. Van Es bought the coaster in 1975 and sold the ship again to Athos Shipping Inc. in Piraeus as **Athos** in July 1984. In 1986 she became the **Xeni M** of Gialos Shipping Co. Ltd., in Limassol.*
*In 1988 Vrahos Navigation Ltd., in Limassol became the new owners and the ship was renamed **Katerina II**. In 1992 she became the **Peter K** of Nief Shipping Co. Ltd., in La Valletta. In 1994 Bernard Arbert Avril in Belize became the new owner and the ship was renamed **God is Love**. Later on Avril Shipping in the Dominican Republic became the owners. She was laid up on 13 February 1995 after a fire had broken out on the ship. In May 1995 she was sold to local shipbreakers in Puerto de Santa Maria.*
Both ships had two cargo holds and two derricks of 4 tons and two of 10 tons.

*Left page above: The **Breehees** in the English Channel.*
*Left page below: The **Breehees** on the New Waterway.*

*Above: The **Breehees** as **Katerina II** riding at anchor in English waters.*
*Middle: The **Breehorn** in English waters.*
Below: In the English Channel.

Next page: General plan of the ships.

P. A. VAN ES & CO. B.V. - ROTTERDAM

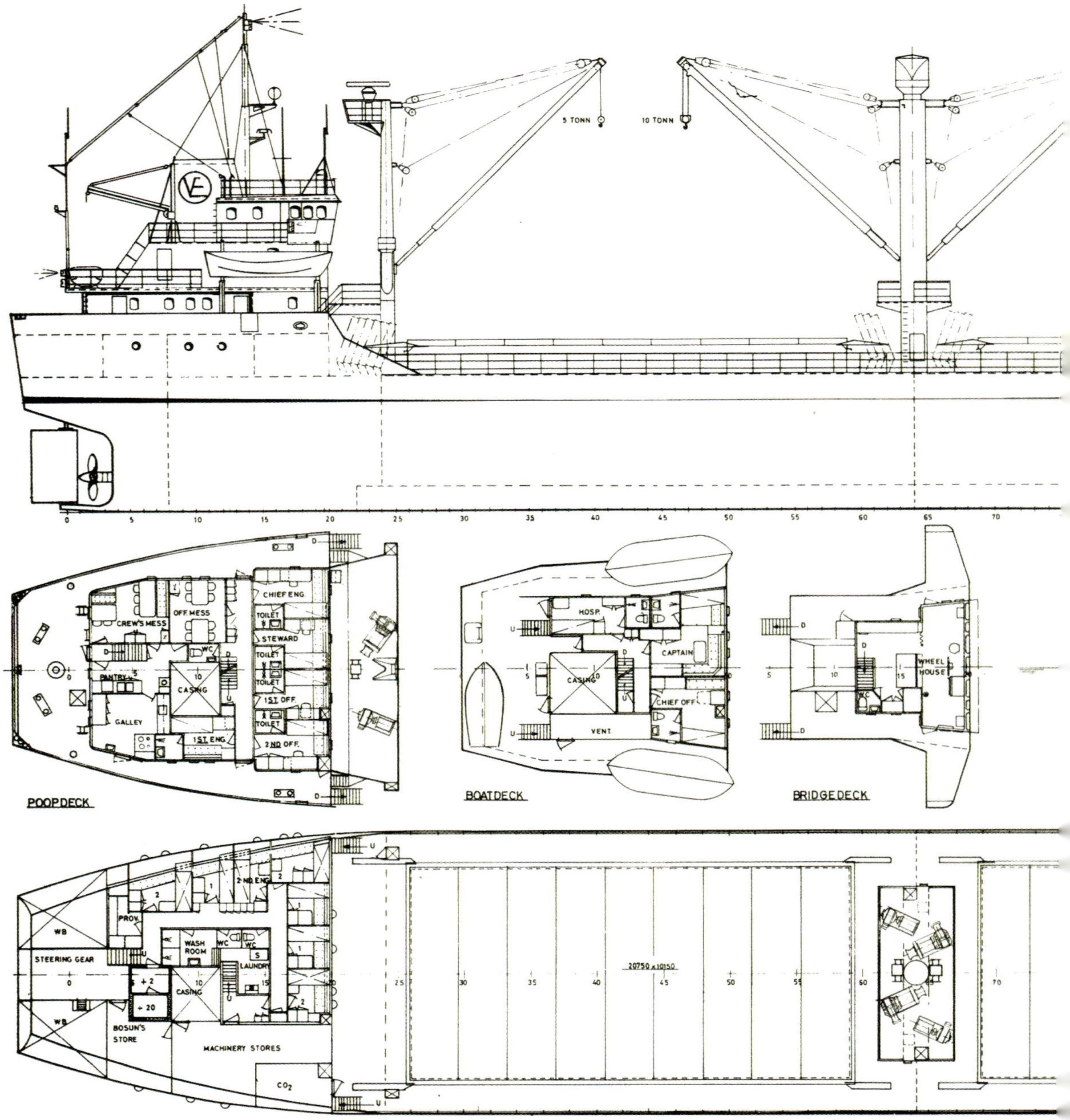

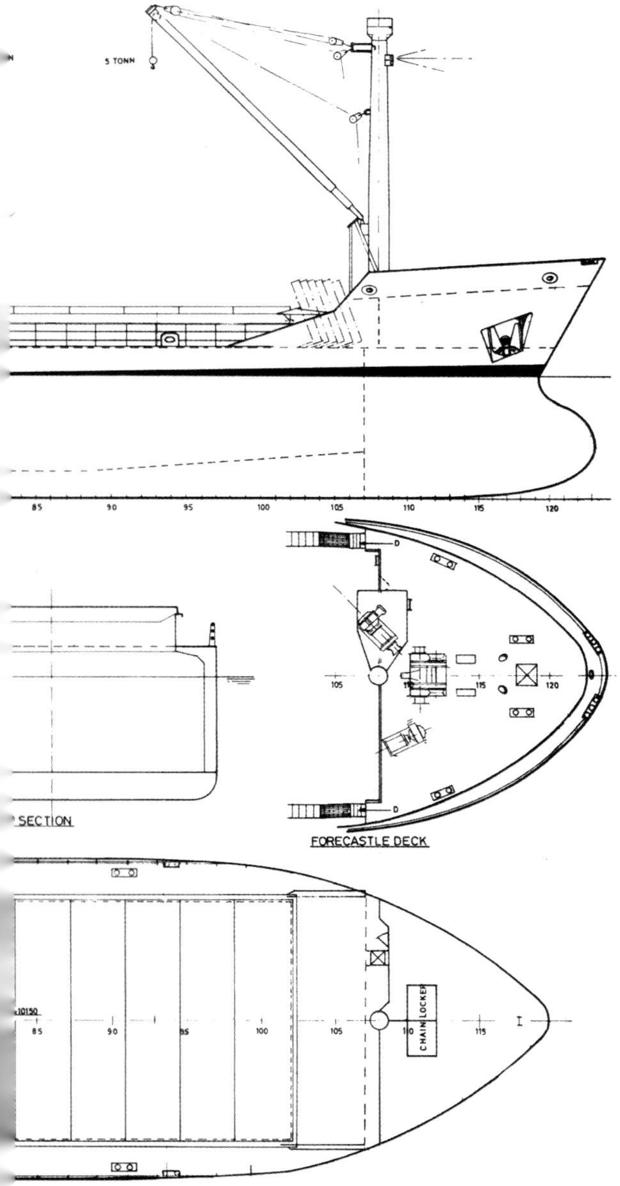

m.v. BREEHORN, Built 1971 -
m.v. BREEHEES, Built 1972

Classed: Det Norske Veritas
✚ 1A1 IS"B" EO

Lenght o.a.	79,98 m	262'	$4^{3}/_{4}$"
Lenght b.p.	74,50 m	244'	5 "
Breadth	13,75 m	45'	$1^{3}/_{8}$"
Depth	6,52 m	21'	$4^{5}/_{8}$"
Draft summer	5,315 m	17'	5 "

Gross tons 1585,78 - 1585,96 RT

Net tons 1117,55 - 1119,07 RT

Deadweight a.t. summer 3017 metric tons

2 Holds with single pull Mac Gregor steel hatch covers

Hatches each hold
20,76 x 10,15 m
68' $1^{3}/_{8}$" x 33' $3^{5}/_{8}$"

Grain fitted : one transverse bulkhead
Breehorn in hold I frame 71
Breehees in hold II frame 55

Grain capacity : hold I 69744 cft
hold II 69638 cft
total 139382 cft

Bale capacity : hold I 63666 cft
hold II 64224 cft
total 127890 cft

Timber capacity: abt. 3800 m^3 truck packages

Derricks : hatch I 1 x 5 + 1 x 5/10 tons
hatch II 1 x 5/10 + 1 x 5 tons
"swinging system"

Hydraulic winches
Water ballast : max. 745 m^3
Bunkers : max. 247 m^3
Engine : 2100 HP
Speed : abt. $12^{1}/_{2}$ knots
Consumption : abt. 8 metric tons per 24 hrs

Radar and VHF fitted

After trials on 16 June 1971, Bodewes Scheepswerven N.V. at Martenshoek delivered the ***Jettie*** *to shipowner Rederij Motorschip 'Jettie' in Zaandam. The management of the 81.41 metres long and 11.90 metres wide ship was consigned to Scheepvaartkantoor J. Tavenier N.V. in Zaandam. At that moment it was the largest short-haul trade ship registered in the Netherlands in the dry cargo sector. It did not sail long for* ***Tavenier****, on 4 August 1972 the* ***Jettie****, measuring 2,854 tons, was sold to C.V. Scheepvaartonderneming 'Klarenbeek' at Willemstad in Curaçao (Spliethoff's Bevrachtingskantoor B.V.) and renamed* ***Prinsenbeek*** *with Willemstad as her home-port. The coaster, equipped with four 5 ton-s derricks, was sold to shipowner H&P (Hendrik & Pieter) Holwerda in Heerenveen on 23 March 1979 and renamed* ***Prinsenwal****. However, on 13 November 1980 she was sold already again to P.T. Pengembangan Armada Niaga Nasional (PANN) (National Fleet Development Corp) in Jakarta and renamed* ***Niaga XXXI****. Keeping her name she was resold in 1992 to P.T. (Persero) Pann Multi Finance in Jakarta. In 1996 P.T. Bahari Haluan Samudra in Jakarta became the new owner and the name* ***Tamara*** *appeared on the bow. Presently, known by this same name the ship sails for Berlian Tirta in Surabaya.*

Above: An official photo of the shipyard.
Middle: The ***Jettie*** *in the summer of 1971 on the North Sea Canal near Assendelft.*
Below: Side-view of the ship.

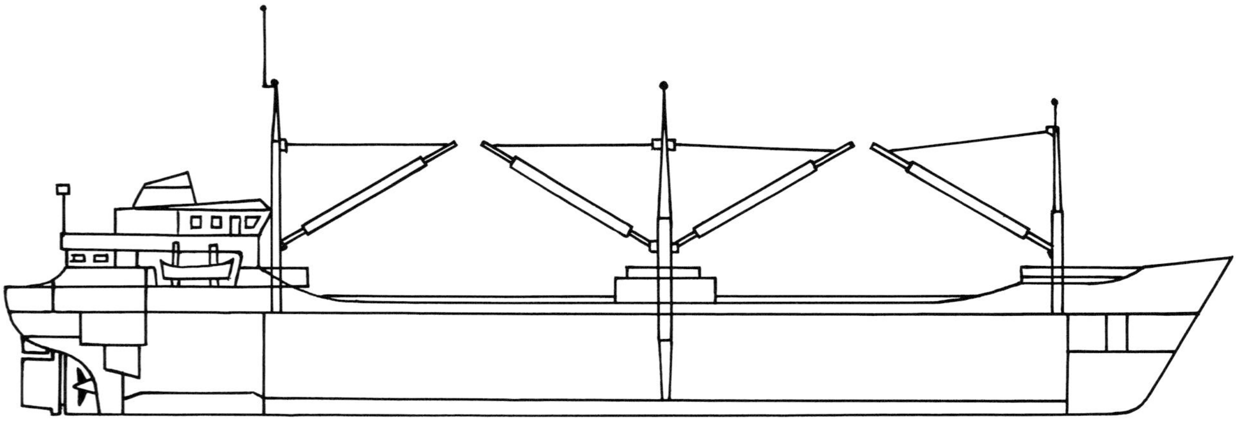

An elegant coaster that reminds me of German building in respect of design but happens to be a true Dutch product was the ***Caravelle*** *of Cebo Marine B.V. in Heemstede. The container coaster was launched as* ***Geestroom*** *at N.V. Scheepsbouwwerf v/h De Groot & Van Vliet in Slikkerveer on 26 November 1971 and she was delivered to Waling van Geest & Zn N.V. in 's Gravenzande on 17 February 1972. In February 1976 the sale of the 70.55 metres long and 10.25 metres wide coaster to A.J.&H. Boeree in Delfzijl followed, who renamed the ship* ***Caravelle****. In February 1989 the ship, which measured 994 tons, was sold under her same name to Cebo Marine B.V. in Heemstede and she was rebuilt at the Oranjewerf in Amsterdam into a cement tanker. The coaster regularly sailed to England and Denmark. After the removal of the cement tanks at the Oranjewerf, the* ***Caravelle*** *was sold to Dick van der Kamp Shipsales B.V. in Brielle on 20 February 2005 whereafter she was laid up on the site of the former shipyard Scheepbouwwerf v/h De Groot & Van Vliet in Slikkerveer. Resold to Lygan Shipping in Panama in March 2005 and towed to Banjul in Gambia by the Danish tug* ***Thor****. The coaster was not allowed to sail by herself because the certificates were expired. On 18 April 2005 her flag became Panamanian. At the moment, the ship still sails with her original name under the flag of the Comoros but further details are unknown.*

Top: Entering IJmuiden on 6 March 2001.
Middle: The ***Caravelle*** *as* ***Geeststroom****.*
Below: The ***Caravelle*** *along the quay at Cebo Marine in IJmuiden on 23 January 2005.*

STAR ARUBA

*The **Star Aruba** was launched as dry cargo coaster **Kröpelin** at the Elbewerft in Boizenburg, Germany on 9 June 1972 and in September of that same year she was delivered to VEB Deutfracht/Seereederei in Rostock, then GDR. The ship, equipped with a bow strengthened for ice fields and originally with two decks and cargo holds, was bought by Koole in Zaandam in 1992 and as **Star Aruba** she was brought in in Tarquin Marine Ltd., in Limassol while Koole kept the management. The 70.97 metres long and 10.37 metres wide coaster was rebuilt into a tanker for the transport of edible oils. At Scheepswerf H. Poppen-Zwartsluis in Zwartsluis ten cargo tanks of 150 m³ were built in.*
*The further completion was done by Koole itself in Zaandam. In April 1997 the **Star Aruba** was reflagged to Dutch flag with Zaandam as her homeport and she was brought in in Star Bonaire B.V. (C. Koole Tanktransport B.V.) in Zaandam. The tanker, measuring 1,531 tons, mainly sails to England and to London in particular.*

*Left: The **Star Aruba** on 7 March 2002 just after leaving the Noordersluis (lock) in IJmuiden, with Zaandam as her homeport. The name is now positioned straight up instead of somewhat curved.*
*Above: The **Star Aruba** on 18 August 1992 on the North Sea Canal before she was rebuilt.*
Below: The coastal tanker on the Humber in England.

The ***Maasborg*** *of E. Wagenborg's Scheepvaart & Expeditiebedrijf N.V. in Delfzijl was built in two sections. The fore part of the ship was built at Scheepswerf 'Friesland' B.V., C. Amels & Zoon in Lemmer and the after part in the covered shipbuilding hall of the shipyard in Makkum in which the two sections were joined to one piece as well. The launch of the shelter deck ship was on 9 August 1974 and the delivery on 5 October 1974. The* ***Maasborg*** *that was put in service for the transport of timber, was 81.82 metres long and 15.27 metres wide. The ship measured 3,658 tons and she had three hydraulic cranes of 8 tons each. On 17 November 1989 the* ***Maasborg*** *was sold to H.J. Smith in Posterholt, without changing her name, she kept sailing for Wagenborg though. On 18 November 2004 she was sold to Asoka Shipping Ltd., in Phnom Penh and she was renamed in* ***Silver Hope****. Presently, she still sails as* ***Silver Hope*** *under Cambodian flag but now for Arnedo Finance Inc. at the Virgin Islands and she is managed by East Marine Co. Ltd., in Vladivostok.*

Above: The ***Maasborg*** *on the North Sea Canal, on 29 October 1997 and page right on 16 February 1998.*

MAASBORG
DELFZIJL
GRANINGE

*A fine coaster is the **Lida** of Lida Shipping B.V. She was launched as **Arina Holwerda** at Barkmeijer Stroobos B.V. in Stroobos on 13 September 1974 and on 1 November 1974 she was delivered to Hendrik and Pieter Holwerda in Heerenveen. She would be managed by Scheepvaartkantoor Holwerda in Heerenveen. They sold the 62.80 metres long and 10.80 metres wide coaster in December 1980 to Femme de Jong in Heerenveen, who named her **Spray**. She kept this name when Scheepvaart Maatschappij Daan in Rotterdam took her over in 1985. In April 1986 C.V. Scheepvaart-onderneming Vissersbank (Pot Scheepvaart) in Delfzijl became the new owner and the coaster, measuring 1,482 tons, was renamed **Vissersbank**. Finally, Jan Wind Shipping in Nansum became the new owner in June 1990. After numerous changes in names of ship-owners and management, the **Lida** presently sails under the flag of Antigua & Barbuda with St. John's as her homeport and the management is dealt whit by Jan Wind Shipping B.V. in Alkmaar. The coaster was completely renovated in 1998/1999 on a former Soviet naval shipyard in Tallinn in Estonia and she was given an accommodation that would envy many a luxurious yacht.*
*The **Lida** is one of few coasters that still have their own loading gear (2 derricks of 2.5 tons each), she sails worldwide and usually with project cargo and bulk cargo.*

*Above: The **Lida** as **Arina Holwerda** during her trial trip on 1 November 1974.*
Middle: Along the Noorderkade in Alkmaar on 7 March 2004.
*Below: The **Lida** moored along the Kanaalkade in Alkmaar on 22 February 2004.*

Scheepswerf & Machinefabriek De Groot & Van Vliet B.V. in Slikkerveer is the shipyard that built the ***Kirsten Smits****. The laying of the keel took place on 19 December 1975 and the launch was on 2 April 1976. The delivery to the Dutch shipowner Marinus Smits, who had his registered office in Vedbaek in Denmark though, was on 26 May 1976. At first, the 84.21 metres long and 14.46 metres wide ship was chartered by Kamp's Scheepvaart- & Handel Maatschappij N.V. in Rotterdam, but soon General Shipping & Chartering Services 'Genchart' VoF took over the management of the ship. The* ***Kirsten Smits****, measuring 3,676 tons, had one cargo hold and she was equipped with a wheeled crane with two jibs that were able to hoist 30 tons in tandem. On 24 November 1983 she was sold to West Indies Shipping Corp in Montserrat and renamed* ***Caricom Express****. Many names and owners would follow after that. In June 2007 she sailed as* ***Princess Ita*** *for Holby Maritime Corporation in Las Palmas at the Canary Islands.*

The photos show the ***Kirsten Smits*** *leaving the Noordersluis (lock) in IJmuiden on her way to Amsterdam on 25 September 1983.*

Brochure published by the shipyard with the technical specifications of the R-type vessels.

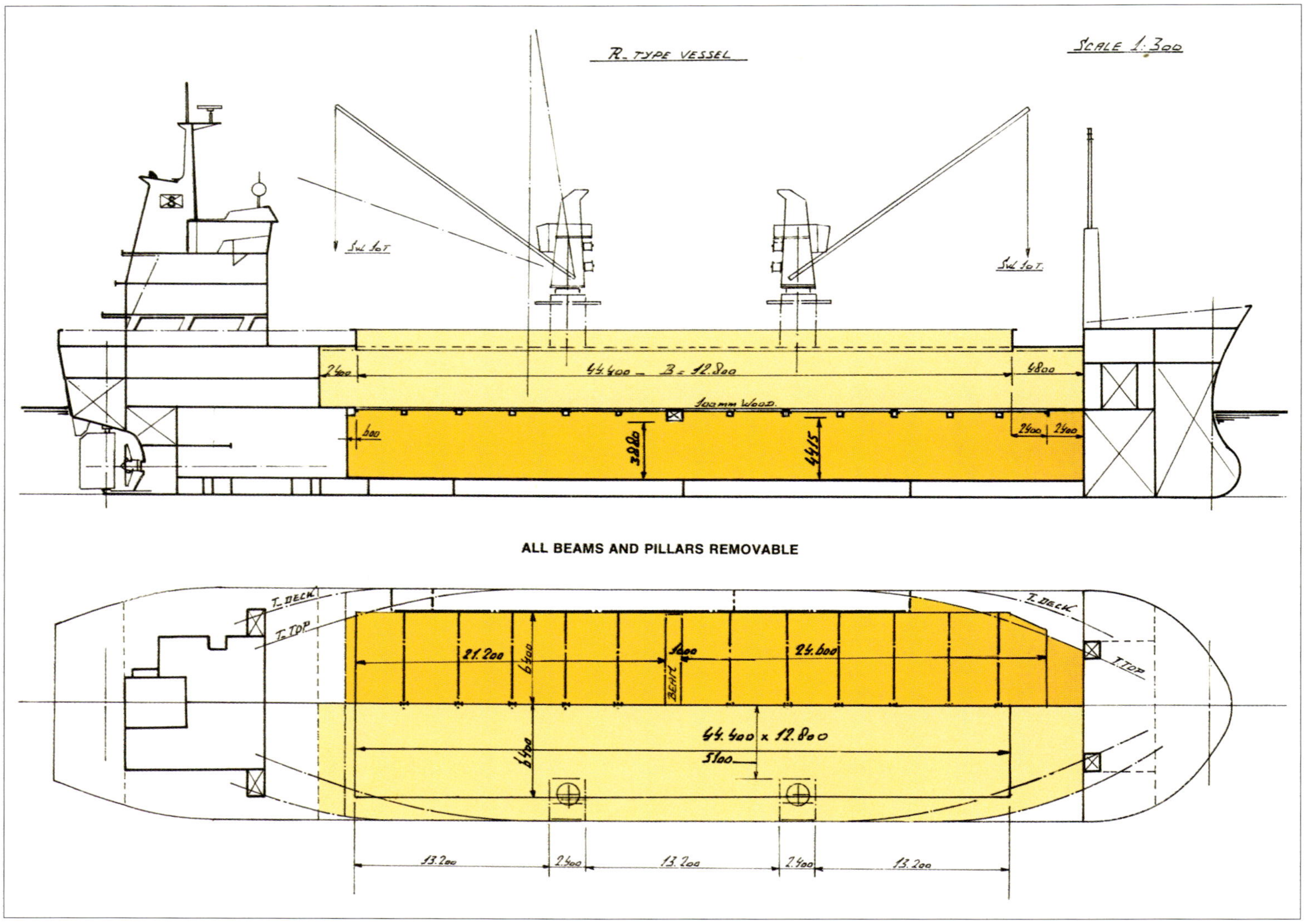

*The **Realengracht** of Scheepvaartonderneming 'Raamgracht' in Amsterdam was launched at Miho Zosensho Shipyard Co. Ltd., in Miho, Shimizu, Japan on 25 October 1976. The delivery followed on 6 December 1976 and the ship was added to the fleet of Spliethoff's Bevrachtingskantoor B.V. in Amsterdam. The 80.22 metres long and 16.03 metres wide ship had a gross tonnage of 1,599 and a deadweight capacity of 3,520 tons as an open shelter deck ship and a gross tonnage of 3,420 and a deadweight capacity of 5,720 as a closed shelter deck ship.*

*In the **Realengracht** stood a 6-cylinder 4S. C.S.A. Hanshin Nainenki Kogyo diesel engine of 3,000 bhp. The **Realengracht** had two deck cranes of 10 tons each.*

*The photo shows the **Realengracht** along the Noorderkade in the port of De Pijp in Beverwijk while discharging a lading of sawed timber from Finland on 4 August 1987.*

Carebeka B.V. in Groningen had the ***Carebeka IX*** *built at Scheepswerf 'Friesland' B.V. in Lemmer.*
The keel of yard number 360 was laid on 27 July 1976 and the launch was on 7 January 1977. The delivery to Rederij 'Carebeka' was on 2 March 1977. The 81.86 metres long and 13.92 metres wide schip had a gross tonnage of 1,598 and a deadweight capacity of 3,120 tons. The ship had a 12-cylinder 4S. C.S.A. Nohab Polar AB diesel engine of 2,640 bhp, enough for 12.5 knots. Due to the very bad financial situation of Carebeka, the ***Carebeka IX*** *was sold to Linde Lloyd B.V. (Gerard van den Bergh) in Ursum and renamed* ***Elsborg*** *on 30 December 1982.*
It came in bare-boat charter for Wagenborg Scheepvaart B.V. in Delfzijl. From 1 April 1984 to 28 October 1986 the ship was managed by Sandfirden B.V. in Haren.
On 17 November 1988 Linde Lloyd sold the coaster to Artemis Marine Ltd. (Vassilios Siskopoulos Marine Enterprises, Thessaloniki), in La Valletta at Malta. She was renamed ***Artemis****. After this it would be renamed several times.*

Above: As ***Carebeka IX*** *in an English port.*
Below: An empty ***Elsborg*** *heading for open sea between the piers of IJmuiden on 16 June 1987.*

*The **North Sea** of Faerder Transport B.V. was built as **Sylvia Beta** at Scheepswerf Niestern in Appingedam and delivered to Sylvia Cargo B.V. in Harlingen on 9 January 1978. Mr. Harm Roossien and others were the people behind this shipowner, who would put many ships into service. However, in March 1985 Sylvia Cargo went bankrupt and from sheer necessity he had to sell the **Sylvia Beta**. On 17 April 1985 the new owner became K.V.O. Elise B.V. (50% O. Muller and 50% H.H.F. Niemeijer) in Groningen and the coaster was renamed **Elise**. After survey, on 14 November 1986 Elise Shipping B.V. in Willemstad, Netherlands Antilles became the new owner. The 81.72 metres long and 14.08 metres wide coaster was managed by Sandfirden B.V. in Haren as from 17 April 1985. In April 1992 the **Elise**, which had a Gt of 1,599 and a deadweight capacity of 3,214 tons, was renamed **North Sea** and the new owner became Faerder Transport B.V in Willemstad. (Bulker Investment Fund BIF, Otto Muller, Hanno & Co, H.H.F. Niemeijer, Ned Investerings-bank NIB). The management remained at Sandfirden until the end of 1993. On 1 January 1994 the management was handed over to Hanno-OAM Shipping B.V. in Willemstad. (H. Hanno & Co B.V.) and Willemstad became the **North Sea**'s homeport.*
*In 1998, Rotterdam became her homeport. In November 2001 Rederij 'Combi Trader' (Hanno Shipping) became the new owner. As from May 1999, Hanno Shipping was a daughter company of Arklow Shipping in Rotterdam, which later became Arklow Shipping Nederland B.V. On 11 February 2001 the coaster was sold to Dolphin Shipping & Services in Phnom Penh in Cambodia and renamed **Farouk J**.*
The present owner is unknown, however, it is said the ship is managed by Harmoush Shipping Co. in Tartus in Syria.

*Above: The **North Sea** in the Noordersluis (lock) in IJmuiden on 6 April 2000.*
Below: After having left the Noordersluis on the same date.

*The **Parkhaven** was launched at Schiffswerft Heinrich Brand K.G. in Oldeburg, Germany on 2 June 1977 and on 25 June she was delivered to Schepers & Co K.G. (See- und Fluss-Transport Gesellschaft mbH, Bremen) in Elsfleth as **Marlene-S**. Bought by N.V. Gebr. Van Uden's Scheepvaart- en Agentuur Maatschappij in Rotterdam on 13 December 1985 and brought in in Rederij 'Parkhaven' V.oF. (Van Uden & G.A. Krikken). Put into service in the Vanuden Africa Line. In January 1993 she was renamed **Umag Saint Malo**, in charter for Armadores Union Maracaibo Aragua Guanco (Van Uden), Willemstad, Netherlands Antilles. In November 1996 she was brought in in Scheepvaart en Handelsonderneming G.A. Krikken B.V. (Noordwest Shipmanagement) in Tynaarlo and renamed **Rodach**.*

*Left page: The superstructure of the **Parkhaven**, IJmuiden, 27 February 1988.*
*Above and middle: The **Parkhaven** leaves the Noordersluis (lock) in IJmuiden on 22 October 1990.*
*Below: The **Parkhaven** in 1994 as **Umag Saint Malo** on the Mersey near Liverpool in charter for the Venezuelan Armadores Union Maracaibo Aragua Guanco.*

On 4 April 1986, Leibulk B.V. of Amsterdam bought the shelter deck ship ***Avebe Star*** *from Rederiet Sverre Leiren A/S of Nesttun, Bergen in Norway. The* ***Avebe Star****, the former* ***Star*** *was built at Soby Motorfabrik og Staalskibsvaerft in Soby, Denmark. She was launched in February 1978 and delivered to P/R Seatown in Soby on 19 May 1978 as* ***Star****, after which she would change owners twice but she would keep the name* ***Star****. In 1986 the 62.66 metres long and 11.23 metres wide coaster was fitted for the transport of flour products; amongst others the after-mast was removed. She had two cargo holds, one of 730 m³ and one of 800 m³, in order to be able to carry different ladings simultaneously. With the coaster, measuring 1,080 tons, Leibulk was going to transport potato starch for Avebe (Aardappel Verkoop Bureau/Potato Sales Agency) in Veendam. This also explains the name* ***Avebe Star****. Annually, 50,000 tons of potato starch was transported from Jonker Veendam in Farmsum to the Humber district and Rochester in England. Due to competition from England this decreased to 30,000 tons. In September 1997 the coaster was renamed* ***Frima Star*** *and all crew matters were dealt with by Wagenborg in Delfzijl.*

On the photo above we see the ***Avebe Star*** *in the Ems Canal moored at Jonker Veendam in Farmsum mid 1986. Below: Painted green in the meantime, we see the* ***Avebe Star*** *on the Ems in front of Delfzijl.*

*The **Inca** was launched at B.V. Nieuwe Noord-Nederlandsche Scheepswerven in Groningen on 12 April 1978 and in May she was delivered to Zeerederij Holland-Friesland B.V. in Rotterdam. The 83.93 metres long and 14.46 metres wide ship was managed by Dammers & van der Heide's Shipping & Trading Co. B.V. in Rotterdam. In 1983 the reefer was lengthened at Nederlandsche Scheepsbouw Maatschappij VoF in Amsterdam. Her length now became 101.94 metres and the gross tonnage was raised from 1,195 to 2,989. The deadweight capacity raised from 2,526 to 3,502 tons. On 22 March 1989 she was consigned to the management of Seatrade Groningen B.V, whereafter she was sold to Caribbean Bulk Carriers B.V. in Willemstad, Netherlands Antilles in January 1991. The management remained at Seatrade. In June 1991 the ship's flag became a flag of convenience, Cyprus, and the new owner Interice Marine Co. Ltd., in Limassol. She came in charter for Thien & Heyenga Bereederungs- und Befrachtungsges. mbH in Hamburg. Seatrade kept the management. On 14 April 2004 the **Inca** was sold to United Gainer Ltd., in Hong Kong. She was renamed **Semper Fidelis** and Phnom Penh in Cambodia became her homeport. The management was handed over to Strofcom Shipping Co. Ltd., in Busan, South Korea. On 4 November 2005 the reefer was renamed **Sun** and on 20 February 2006 **Armont** appeared on her bow.*

*Above: The **Inca** before lengthening.*
Below left: After the lengthening, along the Trawlerkade in the Vissershaven (Fishermen's Port) in IJmuiden on 26 May 1988, loading horse mackerel for Nigeria. The funnel still carried the colours of Dammers.
Below right: Along the Kotterkade in the Haringhaven (Herring Port) in IJmuiden on 6 July 2000. The funnel has the colours of Seatrade and her homeport is Limassol.

In June 1978, Van Goor's Scheepwerf en Machinefabriek in Monnickendam delivered the ***Almenum*** *to Bart & Rienk Switynck in Harlingen. The 63.00 metres long and 9.43 metres wide flush deck ship was temporarily renamed* ***Almenum I*** *in 1992 because the Switynck Bros. were having built a new, larger* ***Almenum****.*
After delivery, the ***Almenum I*** *was sold to Jan de Koning Gans Sr in Hasselt who renamed the coaster* ***Zwartewater****. In September 1995 he, for his part, sold the ship to his son Jan jr who renamed the coaster, measuring 1,114 tons,* ***Oostzee*** *and who transported lots of grain with her from Denmark to the Netherlands. On 16 January 2004 he sold his ship, without renaming her, to Rederij Wantij B.V. in Zwartsluis, established in April 1998 and captains/owners J. van Veen, J. Hartman and N. Blokzijl being the men behind it. They mainly transported steel and bulk with the ship in the North Sea area. The small ship that was able to sail on the rivers in the Netherlands, Belgium and Germany was sold to a shipowner in Riga in Estonia in October 2005, keeping the same name.*

Above: Leaving the port of Rotterdam.
Underneath: As ***Almenum I*** *in Harlingen in May 1992.*
Middle: As ***Zwartewater*** *on the North Sea Canal.*
Below: As ***Oostzee*** *in the 3rd Rijksbinnen-haven in Velsen-Noord on 25 February 2001.*

*The **Leuvehaven**, dating back to June 1978, was bought by N.V. Gebr. Van Uden's Scheepvaart- en Agentuur Maatschappij in Rotterdam in May 1993, as **Snoekgracht** of C.V. Scheepvaartonderneming 'Lauriergracht en Roco' (Spliethoff's Bevrachtingskantoor B.V.) in Amsterdam and brought in in Salpoort Shipping N.V. (Van Uden). Willemstad. The 80.22 metres long and 16.13 metres wide ship was put in service between Kotka (Finland) and Norrköping (Sweden) and Eleusis (Greece) and Istanbul (Turkey). The **Leuvehaven**, measuring 5,895 tons and equipped with two deck cranes of 15 tons each, was chartered to Salpoort Shipping N.V. (Noordwest Shipmanagement B.V., Tynaarlo) in Willemstad, Netherlands Antilles in March 1994 and renamed **Tropical Carrier**. On 1 March 1995 she was sold as **Fiducia**.*

*Above: As **Snoekgracht**.*
*Middle and below: The **Leuvehaven** leaving for open sea on the North Sea Canal on 15 May 1993 during her first voyage for Van Uden.*

MOON BIRD

*A robust looking ship was the **Alamak** of the Rotterdam Intercontinental Line (Zeerederij H. Nefkens) of Rotterdam. The keel was laid at B.V. Scheepswerf v/h Jonker & Stans in Hendrik Ido Ambacht on 30 December 1977, the launch followed on 15 July 1978. The trial trip and delivery of the 82.12 metres long and 17.02 metres wide ship were on 18 September 1978. The **Alamak** had two decks and two cargo holds and a gross tonnage of 3,474 and a deadweight capacity of 5,860 tons. Apart from four derricks of 10 tons each, the **Alamak** had a heavy boom of 100 tons that was secured to a gigantic mast. In December 1984 the ship was renamed **Alamar** and Golden Flash Marine Ltd., in Nicosia, Cyprus became the new owner. In 1987 the ship was sold to Quadro Shipping N.V. in Willemstad (J. Poulsen Shipping A/S), Netherlands Antilles and renamed **Black Bird**. In 1990 the name **Protea Trader** appeared on her bow and in that same year she was renamed once again, now into **Moon Bird** of Bantema Shipping N.V. in Willemstad (Poulsen). In 2003 the ship was sold to Mar Runners Shipping Co. Ltd., in Limassol (flag Cambodia) and renamed **Seagrand**. The management is carried out by Grand Marine in Vladivostok.*

*Left page & above: As **Moon Bird** moored along the Noorderkade in the port De Pijp of Beverwijk on 19 December 2002.*
*Right: She was built as **Alamak**.*

*The **Menna** was launched at Scheepswerf 'Voorwaarts' B.V. in Hoogezand on 7 March 1979 and in April 1979 she was delivered to Rederij ms 'Menna' (which included four shareholders) in Delfzijl. She was managed by Wijnne & Barends B.V. in Delfzijl. The single deck ship was put in service for the transport of wood products from the Baltic. The 75.10 metres long and 12.53 metres wide ship had a gross tonnage of 1,599 and a deadweight capacity of 2,828 tons. In the engine room stood a 12-cylinder 2S. C.S.A. Brons diesel engine of 1,500 bhp. On 6 June 1995 the **Menna** was sold to Rederij 'Sanitas' (which means health in Latin) in Delfzijl. The man behind this shipowner was Mr. Willem Hoogenhout, who renamed the coaster **Erna**, the name of his wife. The ship remained in the management of Wijnne & Barends. On 11 January 2002 he sold the **Erna** to Torso Rederi AB in Lidköping in Sweden. She was handed over on 25 February 2005. The coaster got a blue hull and she was renamed **Windena**. Presently she is managed by O.P. Svenssons in Gothenburg.*

*Above: The **Menna** at Delfzijl in August 1981.*
*Middle and under: The **Erna** arriving in IJmuiden with a lading of timber on 7 October 2001.*

The eighties

*On the 20th of May 1987, Gebr. Van Uden's Scheepvaart- en Agentuur Maatschappij N.V. in Rotterdam bought the **Breslau** from m.s. 'Breslau' Holnis Reederei GmbH & Co. in Panama (Hanseatic Shipping Company Ltd., in Limassol) and renamed her **Maashaven**. She had been launched as **Breslau** at Flensburger Schiffsbau-Gesellschaft in Flensburg on 15 November 1979 and due to a three-year time-charter she was delivered to the Dutch Jeco Line on 12 February 1980 as **Togo Brewer**. The 117.82 metres long and 18.54 metres wide ship was renamed **Jomsborg** by Van Uden in 1988 because of a charter to Danneborg Rederi A/B in Copenhagen. In December 1989 she was renamed **Maashaven** again and put back in service on the Vanuden Africa Line.*
*The **Maashaven**, having a gross tonnage of 4,015 and a deadweight capacity of 7,550 tons, had three cargo holds and three tween-decks. The loading gear consisted of four derricks which could lift between 12 and 24 tons.*
*In March 1994 the **Maashaven** was withdrawn from the Africa trade and her flag became a flag of convenience, of the Netherlands Antilles. She was renamed **Aruba** with owners Ivoros Shipping N.V. in Willemstad. Van Uden kept the management and was co-owner.*

*The **Maashaven** on her way to open sea, sailing on the North Sea Canal on 23 October 1991.*

VHF
3
ALMAR
AJ

*The **Almar** was built as **Birgit Jürgens** for Klaus Jürgens in Glückstadt. The keel was laid at J.J. Sietas Schiffswerft GmbH & Co. K.G. in Hamburg on 21 May 1980 and on 13 September 1980 the coaster was delivered.*
*The 85.83 metres long and 11.33 metres wide ship was sold to Tantomar Transportes Maritimos Ltda in Lisbon on 13 September 1988, who renamed her **Almar**. In 1994 she was sold to Slibail Portuguesa de Locaçao Financiera S.A. in Lisbon (management by Tantomar). In 1997 Woodchester Credit Lyonnais Leasing S.A. in Lisbon (Tantomar) became the owner. On 10 September 1997 the ship arrives in Rotterdam and on 3 November 1997 she is apprehended, whereafter she is laid up under arrest in Slikkerveer. In November 1999 A.J. van der Kooy B.V. in Pijnacker takes over the ship from Dick van der Kamp Shipsales B.V. in Brielle. The crawl-line coaster is then rebuilt at Vlaardingen Oost Shiprepair B.V. in Vlaardingen (built-in tanks) to a tanker for the transport of edible oils. The rebuilding is completed at Scheepswerf Schiepo in Ridderkerk on their own account. On 9 March 2001 the ship is christened by the shipowner's daughter, Ms. Annette van der Kooy. Next, on 13 March 2001, she is registered in Willemstad, Netherlands Antilles and now her new owner is Scheepvaart Maatschappij 'Almar' B.V. The management remains at Kooy Shipping B.V.*
*As per 1 January 2007 new, more stringent regulations for waste from ships have become effective: Wet Voorkoming Verontreiniging door Schepen (WVVS) (Law Preventing Pollution by Ships). This stems from new Marpol regulations (International Convention for the Prevention of Pollution from Ships). Did the **Almar** previously sail to 17 countries, due to the new regulations the ship is only allowed to sail from the Netherlands to Belgium, France, Germany and England.*

*Left: The **Almar** in the Middensluis (lock) in IJmuiden on 25 January 2002.*
*Above: The **Almar** sailing on the North Sea Canal on 13 February 2003. She has a Gt of 1,698 and a deadweight capacity of 2,208 tons.*
Below: The tanker leaves the Middensluis (lock) in IJmuiden for open sea on 24 August 2002.

*The **Bickersgracht** was launched at Miho Zosensho K.K. in Shimizu, Japan on 17 December 1980 and after trials on the 13th of January, she was delivered to C.V. Scheepvaart-onderneming 'Brouwersgracht' on 6 February 1981.*
The ship's length was 80.22 metres and her breadth 16.11 metres. Her gross tonnage was 1,599 and the deadweight capacity 3,488 as an open shelter deck ship and 3,467 and 6,195 tons respectively as a closed shelter deck ship.
*The **Bakengracht** was launched at the same shipyard on 22 January 1981 and she was delivered to C.V. Scheepvaartonderneming 'Bakengracht' on 17 March 1981.*
*The ships were managed by Spliethoff's Bevrachtings-kantoor B.V. in Amsterdam. The dimensions and tonnage of the **Bakengracht** were almost identical to those of the **Bickersgracht**. Both ships had a 6-cylinder 4S. C.S.A. Hanshin Nainenki Kogyo diesel engine of 3,200 bhp. They had one cargo hold and a tween-deck as well as two deck cranes of 15 tons each.*
*The **Bickersgracht** was sold to Deepocean Shipping N.V. in Willemstad, Netherlands Antilles (Normed International B.V.) in February 1999 and renamed **Valmin**.*
*On 12 December 2001 she was sold to Malta and renamed **Orion B**. Since 21 August 2006 she sails again under the name of **Bickersgracht** for Zodiac Shipping in La Valletta and she is managed by Nordost Shipping Ltd., in Petrozavodsk in Russia.*
*The **Bakengracht** was sold to Malta as **Kestrel** on 5 February 1999. On 19 February 2004 she was renamed **Sea Tiger** and on 12 July 2004 renamed again in **Kestrel.** Following various owners she presently sails as **Seatiger** under Cambodian flag for Delouna Shipping in Limassol and she is managed by Grand Marine in Vladivostok.*

*Above: The **Bickersgracht** just after leaving the Middensluis (lock) in IJmuiden on her way to Amsterdam on 10 March 1996.*
*Left: The **Bakengracht** leaves IJmuiden in ballast for open sea on 21 August 1987.*

North Sea Petroleum B.V., the permanent bunker supplier of Wagenborg was the owner of the ***Polarborg****. The ship was launched at B.V. Nieuwe Noord-Nederlandsche Scheepswerven in Groningen on 27 March 1982 and after trials on 18 May she was delivered to Butraco B.V. in Groningen on 22 May 1982. The 82.37 metres long and 15.83 metres wide coaster was brought in in bare-boat charter to Wagenborg Scheepvaart B.V. in Delfzijl.*
The ***Polarborg****, which caught the eye with her two funnels, had a gross tonnage of 1,600 and a deadweight capacity of 3,760 tons.*
On 2 June 1995 the ship was sold to C.V. N&F Bruins (Nicolaas & Filip) in Hardinxveld-Giessendam and renamed ***Hendrik B****.*
She remained managed by Wagenborg. On 27 May 2002 she was sold as ***Arkaim III*** *to Cambodia. At the moment the ship still sails under this same name under Cambodian flag for Terena on St Kitts & Nevis and she is managed by Arkaim in Vanino in Russia.*

Above: The ***Polarborg*** *trying to find her way through the ice, loaded with quite some deck load.*
Below: The ***Hendrik B*** *moored in the Middensluis (lock) in IJmuiden on 18 October 1998.*

*The flush deck ship **Liamare** was launched at Peters Scheepsbouw N.V. in Kampen on 27 October 1982. Trials took place on the 28th of December 1982 after which the ship would be delivered. However, due to overdue paint-work the delivery was postponed for a week. On 5 January 1983 she was put into service as **Dependent** for J. Visser and Gerard J. Schalk in IJsselmuiden. In January 1996 Gerard J. Schalk and Peter Schalk became the single owners of the 65.51 metres long and 11.36 metres wide coaster. They sailed the ship until December 2000 where-upon Scheepvaartbedrijf A.J. Kok in Delfzijl became the new owner and they renamed her **Liamare**. The gross ton-nage of this fine coaster amounts to 1,241 and the dead-weight capacity is 1,791 tons. The ship has a 6-cylinder 4S. C.S.A. Stork-Werkspoor 6FHD240 diesel engine of 1,080 bhp, enough for a speed of 11 knots. On 28 April 2004 Vink Scheepvaart B.V. (T.H. Vink) in Delfzijl became the new owner and she was given the name **Henny**. Ancora Shipping B.V., nowadays Flinter Shipping B.V. in Rotterdam is the ship's shipbroker.*

*The **Liamare** in the Middensluis in IJmuiden on 3 Oct. 2001.*

*The **Vrouwe Johanna** was a real fine coaster. The keel with yard number 615 was laid at Scheepswerf Bijlholt B.V. in Foxhol in June 1982. The launch was on 9 July 1983 and after trials on 8 October 1983 delivery followed to Rederij 'Varend Vervoer' B.V. in Rotterdam. The 81.72 metres long and 14.10 metres wide ship would be managed by Visser & Visser Chartering B.V. in Vlaardingen. The coaster had a gross tonnage of 1,599 (at present 1,980) and a deadweight capacity of 3,008 tons. A 6-cylinder 4S. C.S.A. Caterpillar, type 3606 diesel engine of 1,801 bhp provides the propulsion. The coaster has one cargo hold of 53.30 metres in length and two deck cranes of 12.5 tons each. Room is available for 95 TEU. In 1995 the ship was sold to JT Shipping Ltd., in St John's, Antigua & Barbuda and renamed **Johanna Trader**. She would be managed by Arpa Shipping B.V. in Roosendaal. On 31 October 2002 she is sold to Amazon Shipping Ltd., also residing in St John's and she is renamed **Suzie Q**. She is now managed by Jan Wind Shipping B.V. in Alkmaar.*

*Above: As **Vrouwe Johanna** moored along the Noorderkade in the port De Pijp of Beverwijk on 8 June 1990.*
*Right: As **Suzie Q** along the same quay on 3 May 2007.*

*The **Koningsgracht** of the K1-type was launched at Miho Zosensho K.K. in Shimizu in Japan on 15 July 1983 and on 16 September 1983 she was delivered to C.V. Scheepvaartonderneming 'Koningsgracht' in Amsterdam. The 106.30 metres long and 16.62 metres wide ship was managed by Spliethoff's Bevrachtingskantoor B.V. in Amsterdam. She had a gross tonnage of 2,368 as an open shelter deck ship and 4,983 as a closed one. The deadweight capacity was 4,991 tons as an open shelter deck ship and 8,138 tons as a closed one. In the ship stood a 6-cylinder 4S. C.S.A. Hanshin diesel engine of 3,350 bhp. Room was available for 365 TEU. The **Koningsgracht** had one cargo hold, two hatches and two deck cranes that each could lift 25 tons. On 9 April 2002 she was sold to Corn Marine Ltd., in Istanbul and renamed **Corn Flower**. At the moment Corn Flower in La Valletta is the owner and Cornships Management in Istanbul the manager.*

*Above left: The **Koningsgracht** in dock at Shipdock B.V. in Amsterdam for a large survey on 15 September 1988.*
*Above: The **Koningsgracht** on the North Sea Canal on her way to Shipdock on 13 September 1988.*
Right: Lying in dock, where work is carried out on her propeller and rudder, on 15 September 1988.
*Below left: Sailing with a heavily loaded draught, the **Koningsgracht** enters a port in New Zealand.*

*The keel of the **Unden** was laid on 27 April 1983 and on 10 December 1983 she athwart-ships ran down the slipway at B.V. Nieuwe Noord-Nederlandsche Scheepswerven in Groningen. On 1 February 1984 the delivery took place to Ahlmarco B.V. in Delfzijl. This is a full daughter company of O.F. Ahlmark & Co. Eftr A/B, in Karlstad. The **Unden** was specially designed for service through the Trollhättan Canal in Sweden. The 87.03 metres long and 13.21 metres wide ship would be managed by Sandfirden B.V. in Haren. The **Unden**, with a gross tonnage of 2,610 and a deadweight capacity of 3,171 tons has one cargo hold and a Hägglunds deck crane of 25 tons. Room is available for 135 TEU. In 1993, Forest Line GmbH in Hamburg became the new owner and the Unden's flag became Antigua & Barbudan. The management was now in the hands of Scanunit Marine Management AB. In 1996, Ferm Marine in Willemstad, Netherlands Antilles became the owner and Willemstad her homeport. The ship is managed by Ahlmark Lines AB in Karlstad. In 2007, Barkey Corp N.V. in Willemstad becomes the owner of the **Unden**. As per 19 January 2007 the **Unden** is managed by Marin Ship Management B.V. in Farmsum.*

*The **Unden** sailing on the North Sea Canal on her way to open sea on 12 January 1993.*

The **Dutch Pilot** was built as **Jaqueline Broere** at Scheepswerf Verolme N.V. in Heusden. The launch was on 7 January 1984, trials took place on the 22nd of March and the delivery to Gebr. Broere in Dordrecht was on 23 March 1984. The chemicals tanker is 80.68 metres long and her breadth is 13.70 metres. The gross tonnage is 1,793 and the deadweight capacity 2,760 tons. On the 17th of January 1986 Jaciparanacon B.V. became the owner of the tanker while the management remained at the Broere Bros. In 1993 the tanker is lengthened at Scheepswerf Welgelegen B.V. in Harlingen and also provided with a new double-hulled fore- and midship. The length now is 91.10 metres with a breadth of 13.70 metres. The gross tonnage is 2,137 and the deadweight capacity 3,160 tons. She is relaunched again on 17 July 1993 and then rechristened by Ms. Van der Laan in **Dutch Pilot**. The delivery is 12 August 1993 and the ship is brought in in N.V. Tankvaart Dordrecht in Dordrecht. On 6 June 1994 the new owner becomes Mts Dutch Pilot B.V. in Dordrecht (Gebr.Broere), and thereafter the ship is stored under the umbrella of GB Shipping Services B.V. As from 17 December 1997 Nijverheid B.V. is her new owner. She stays managed by GB Shipping Services B.V. Successively, the ship will be managed by Broere Shipping B.V., Vopak Chemical Tankers B.V. and then in 2005 once again by Broere Shipping B.V. The parent company Broere was taken over and changed names several times. During the Vopak period they established the VCT Holding B.V. This holding was sold to DAL Essberger on 28 April 2004. On that date VCT Holding B.V. transferred to GB Shipping Holding B.V. which still is the Dutch parent company. Since then the main shareholder is the German DAL Essberger group. The official name is Rantzau & Co. Handels- und Beteiligungsgesellschaft GmbH & Co K.G.

Above: Still named **Jacqueline Broere** she is just about to enter the Noordersluis (lock) in IJmuiden.
Below: The **Dutch Pilot** sailing towards sea on the Westerscheldt near the Nauw van Bath (the Straits of Bath) on 14 September 2004.

At Scheepswerf & Machinefabriek De Groot & Van Vliet B.V. in Slikkerveer, the keel was laid for the **Carina Smits** on 26 September 1983. The launch was on 16 March 1984 and trials took place on the 15th of May. Delivery to the Dutch shipowner, but living in Denmark, Marinus Smits was on 17 May 1984. She would be managed by GenChart Vof in Rotterdam. The **Carina Smits** was 113.47 metres long and 17.12 metres wide. The ship, with a gross tonnage of 5,943 and a deadweight capacity of 9,620 tons, had two cargo holds and two hatches as well as three Brattvag deck cranes that were able to lift between 18 and 32 tons. A 6-cylinder 4S. C.S.A. Stork-Werkspoor 6TM410 diesel engine of 4,270 bhp provided a speed of 13 knots. In 1993 she was put under the Cypriot flag of convenience and Rosepearl Shipping Co. in Limassol became the new owner. The **Carina Smits** was now renamed **Carina I** and would now be managed by H.H.S. Holland Ship Service B.V. in Rotterdam. In 1996 the ship was sold to Neptune Investment Ltd., in Port Vila at Vanuatu and renamed **Capitaine Cook**. Onego Pride B.V. (Th. Jacobsen Management AS in Sarpsborg, Norway) became the new owner on 16 August 2004. Sailing under Norwegian flag (NIS) her new name also would be **Onego Pride**. As from 28 November 2006 she flies the Panamanian flag and she sails as **Lady Nour** for Nour Shipping Ltd., being managed by Mamari Shipping Co. in Tartous in Syria.

Above: Riding at anchor in tropical waters.
Below: The **Carina Smits** just before mooring in the Hoogovenhaven at Hoogovens (Corus) in IJmuiden on 31 January 1991.

The **Alstern** was launched on 15 June 1984 as **Friesland** at Zhong Hua Shipyard in Shanghai and after trials on 17 December 1984 transferred to Schenefelder Schiffarts GmbH & Co. ms Hagen K.G. in Hamburg. In 1985 she was renamed **Scol Trader**. The 105.87 metres long and 17.56 metres wide ship had a gross tonnage of 1,597 (4,021) and a deadweight capacity of 4,400 (6,163) tons. The ship had one cargo hold, two hatches and two deck cranes of 25 tons each. The **Scol Trader** was laid up in Hamburg in April 1986 and auctioned in August 1986. The new owner Ahlmark Lines A.B. in Karlstad renamed the ship **Alstern**. On 2 March 1988, **Ahlmark** brings in the ship in Sandfirden B.V. in Haren. The new owner now became in December 1988 Ahlmarco B.V. with Delfzijl as the homeport of the ship. She was put into service for the transport of products from the woods in north-western Europe. After being managed by Scanunit Marine Management AB the **Alstern** was put in 1995 into the management hands of Heinrich Hanno & Co in Rotterdam. In June 1995 she was sold to Barkey Corp (Ahlmark, Karlstad). In january 1997 Ahlmark Lines AB, in Karlstad became the new owner and the **Alstern** was managed by Barber Ship Management A.S.. On the 8th of August 2005 she was transferred to Barkey Corp N.V., Karlstad. The ship sails the Swedish flag. In 2006 Barkey Corp N.V. in Willemstad became her new owner. As per 26 February 2007 she is managed by Marin Ship Management in Farmsum.

Above: On the North Sea Canal on 15 June 1993.
Below: The **Alstern** just after leaving the Noordersluis (lock) in IJmuiden on 21 November 1988.

*The K series of Spliethoff consisted of six ships, four of the K1-type and two of the K2-type. The **Kaapgracht** is a K2-type. Built at the same shipyard (Miho Zosensho K.K.) as the K1-type and launched on 25 June 1984 and delivered to Shipping Carrier Co. N.V. in Amsterdam on the 1st of October 1984. The ships did not differ that much in dimensions and tonnage. However, what did strike was a different construction of the accommodation and the position of the funnel. At the other ships this would be positioned in the middle of the bridge deck together with the radar mast continuing as one entity. At the **Kaapgracht** this was positioned on the port side of the ship with the radar mast detached. The engine capacity differed too, in the **Koningsgracht** described earlier (page 172/173) that was a mere 3,350 bhp, in the **Kaapgracht** and also with a 6-cylinder 4S. C.S.A. Hanshin diesel engine this had been raised to 4,504 bhp. An other difference could be found in the deck cranes, they were able to lift 50 tons instead of 25.*
*On 5 February 2004 the **Kaapgracht** was renamed **Corn Brook** at Shipdock in Amsterdam. The new owner would be Cornships Management Ltd., in Istanbul.*

*The **Kaapgracht** in front and in the Noordersluis (lock) in IJmuiden on 23 February 2002.*

*The **IJsselland** which was launched at B.V. Nieuwe Noord-Nederlandsche Scheepswerven in Groningen on the 17th of November 1984 was supposed to be put in service as **Lauwersborg**. After trials, the ship was delivered to Linde Lloyd Zeeschepen IV B.V. in Zaandam.(Gerard van den Bergh) on 9 January 1985. She was brought in in bare-boat charter to Wagenborg Scheepvaart B.V. in Delfzijl. The 96.78 metres long and 15.43 metres wide ship had a gross tonnage of 3,757 and a deadweight capacity of 5,327 tons. She was a somewhat larger version of the type **Lindeborg**. On 16 July 1996 the ship was renamed **IJsselborg** with C.V. Scheepvaartonderneming 'IJsselborg' in Delfzijl as her new owner. On 16 February 1999 H.J. Smith and S.M. Smith from Delfzijl became the new owners.*
The management remained at Wagenborg.
*On the 21st of February 2005 the **IJsselborg** was sold to Anabatik Ltd., of the Virgin Islands and renamed **Shemut**. She is managed by Natie Shipping Co. Ltd., in Vladivostok.*

*Above: The **IJsselland** in front of the Middensluis (lock) in IJmuiden on 7 June 1988.*
Middle: Moored at the Velser Terminal along the Velser quay in Side Canal A in Velsen-Noord on 26 July 1990.
*Below: As **IJsselborg** she leaves IJmuiden on 16 January 1999.*

*Especially for the transport of steel from Hoogovens (Dutch Blast Furnaces) in IJmuiden (the present Corus) to England, two ships were built. One of them was the **Steel Shuttle** launched at Scheepswerf & Machinefabriek De Groot & Van Vliet B.V. in Slikkerveer (YVC) on 30 August 1985. Trials took place on 25 September and on the 4th of October 1985 the **Steel Shuttle** was delivered to General Shipping & Chartering Services VoF 'GenChart' in Rotterdam. The small coaster had a length of 64.78 metres and a breadth of 11.10 metres. The gross tonnage was 993 and the deadweight capacity 1,715 tons. The engine room housed a 5-cylinder 2S. C.S.A. Bolnes 5DNL190/600 diesel engine of 880 bhp. Amongst others she sailed to Poole on the southern coast of England and Rochester, King's Lynn, Whitby, Seaham, Howdendyke, Goole and Grangemouth on the east coast.*
*In 1993 the **Steel Shuttle** was brought in in Steel Transport B.V. in Rotterdam while the management remained at GenChart.*
*On 17 December 2001 the **Steel Shuttle** was put under Cypriot flag because the transport of steel to England decreased considerably. She was brought in in Omniseas Shipping Company Ltd., in Limassol and renamed **Joanne**. The management was now carried out by HSS Holland Ship Service B.V. in Rotterdam. On the 5th of January 2004 the **Joanne** was laid up in Vlaardingen and put up for sale. In May 2004 Interway Shipping Ltd., in Limassol became the new owner and the coaster was renamed **Mona**. Q-Shipping B.V. in Ridderkerk took care of the management. In August 2006 she was sold to Royal Bengal Shipping Lines in Bangladesh. Her present name and whereabouts are unknown.*

*Above left: The **Steel Shuttle** leaves IJmuiden on 21 August 1997.*
*Middle As **Joanne** in the Middensluis (lock) in IJmuiden on 26 April 2003.*
Below: Laden with steel for England in the Middensluis in IJmuiden on 15 July 2000.

*On 8 August 1986 the **Olivier** was launched at the former premises of B.V. Nieuwe Noord-Nederlandsche Scheepswerven in Groningen. This shipyard had been declared bankrupt on 14 March 1986. Sandfirden in Haren temporarily took over the shipyard and its personnel and thus it happened that the **Olivier** athwart-ships ran down the slipway on the shipyard of Icestar B.V. Trials were on 7 September 1986, followed by the delivery to Seasand C.V. a day later. The ship had been built by order of Erik Thun (Thun Ship Management AB). She was managed by Sandfirden in Haren. The **Olivier** is 88.02 metres long and she has a breadth of 13.17 metres with a gross tonnage of 1,999 and a deadweight capacity of 3,990 tons. On 1 january 1994 the management of the **Olivier** is transferred to Heinrich Hanno & Co. in Rotterdam. In 1995 the new name of the shipowners becomes Rederij Thun II in Delfzijl and the management of the **Olivier** is put into the hands of Marin Ship Management BV in Farmsum. On 18 December 2003 she is sold to Cyprus. Having kept her name, she now sails for Northern Victory Ships in Limassol and she is managed by Bjare Shipping in Båstad in Sweden.*

*The **Olivier** on 22 October 2000 just before sailing in and entering the Noordersluis (lock) in IJmuiden.*

WILLY

The low-profile coaster ***Willy*** *was built as yard number 243 at Scheepswerf Ferus Smit B.V. in Foxhol. She was launched on 8 February 1986 and on 17 March 1986 she was delivered to Houtrak Scheepvaart in Oudeschild, with Evert Bays C.V. in Oudeschild as the owner. The small coaster had a length of 64.29 metres and a breadth of 10.50 metres. The gross tonnage amounted to 863 and the deadweight capacity to 1,281 tons. The engine room housed a 8-cylinder 4S. C.S.A. Caterpillar diesel engine of 775 bhp. On 24 January 2000 and due to failure of the captain (blackout), the* ***Willy****, on her way from Drogheda in Ireland to Kampen with a cargo of 1,150 tons of bone meal, would sail right into the head of the Java quay in Amsterdam. With considerable damages to the foreship as a result. At Frisian Shipyard Welgelegen B.V. in Harlingen the low-profile coaster was repaired in 18 days. Early 2001 the* ***Willy*** *went to Welgelegen once again for a repair job. This was finished in March however, the coaster did not left the shipyard until 2 August 2001. Meanwhile the ship's certificates were expired. On 18 March 2002 Scheepvaartonderneming Anna B.V. in Heerenveen (Holwerda) bought the low-profile coaster and renamed her* ***Anna****. On 12 October 2004 the* ***Anna*** *was sold to Aggregate Carriers Ltd., in Rochester who put her under Bahamian flag as* ***Christine Y****. Manager is Coastal Bulk Shipping Ltd., in Rochester.*

Left page: The ***Willy*** *leaves Hoek van Holland (Hook of Holland) in stormy weather on 19 August 1999.*
Above left: The ***Willy*** *is repaired at Frisian Shipyard Welgelegen B.V. in Harlingen on 10 February 2001.*
Above right: As ***Anna*** *on 23 May 2002 in the North Sea Canal just before sailing in the Zuidersluis (lock) in IJmuiden.*

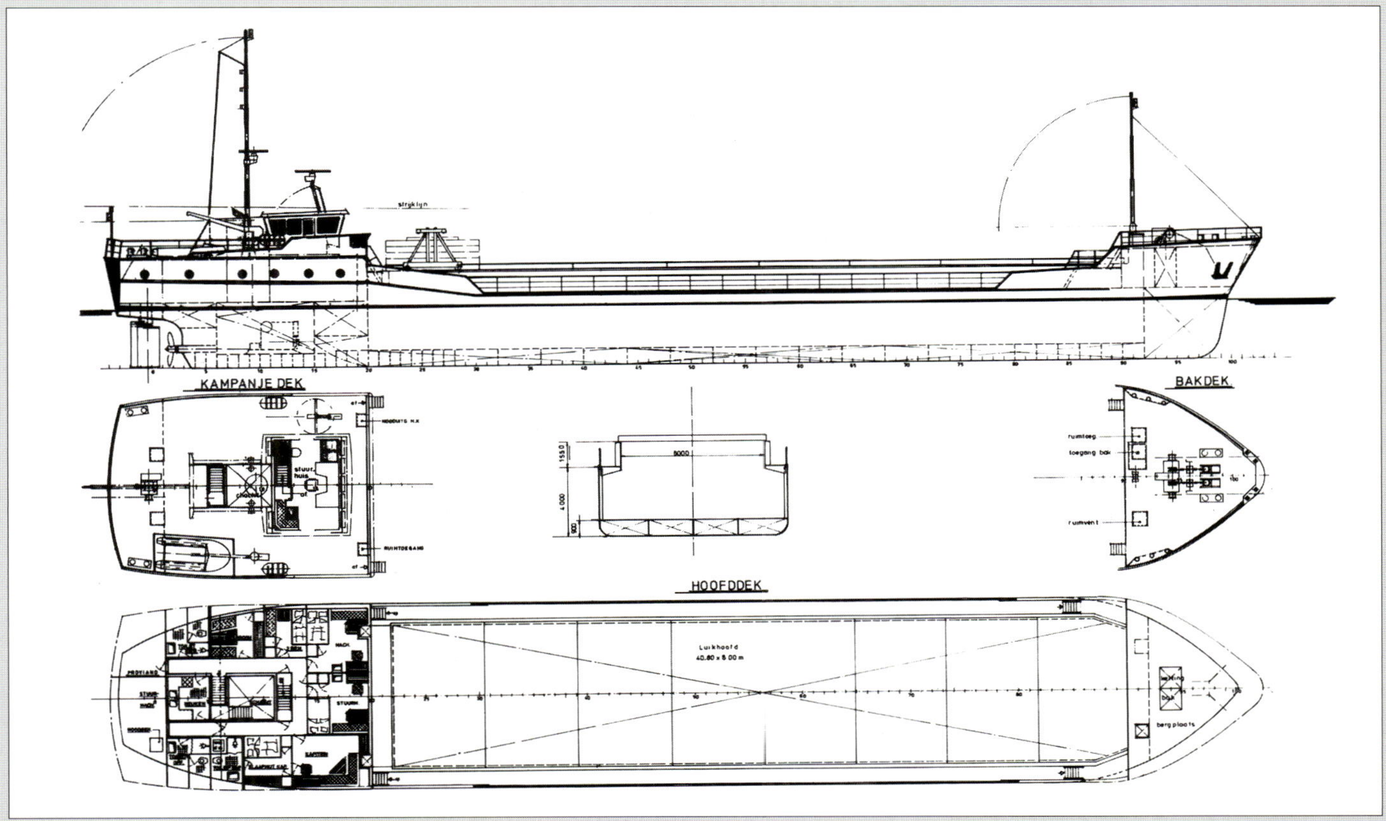

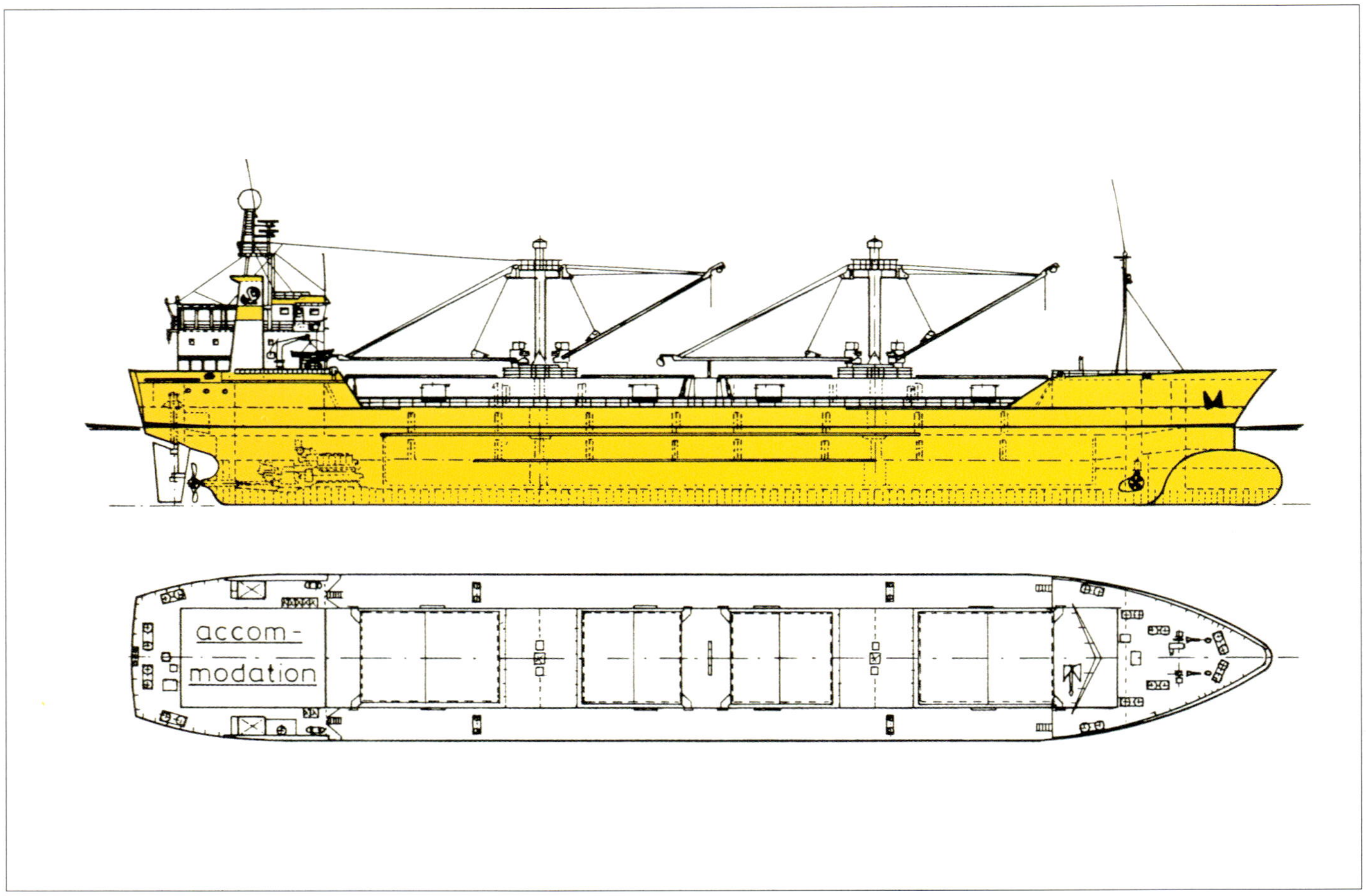

*Below: The **Nostalgic** on 4 October 2000 along the same quay.*

The reefer ***Nostalgic*** *was originally built as* ***Northern Express****. She was launched at B.V. Scheepswerven v/h Gebr. van Diepen in Waterhuizen in August 1986 and after trials she was delivered to Motorschip Northern Express C.V. in Breskens on 7 October 1986. The ship, with a length of 109.07 metres and a breadth of 16.34 metres, would be managed by Vroon B.V. in Breskens. Seatrade Groningen B.V. in Groningen would be the shipbroker of this ship which has a gross tonnage of 3,978 and a deadweight capacity of 5,175 tons. In 1993 the refrigerated ship's flag became a flag of convenience, of Manila, the Philippines to be precise. Ideal Maritime Corp. now became her new owner while Vroon B.V. remained in charge of her management. The reefer has three cargo holds and three decks as well as four Hatlapa mast cranes of 5 tons each. On 22 August 2000 Nostalgic Shipping Co. N.V. in Willemstad, Netherlands Antilles became the new owner and she was renamed* ***Nostalgic****. Seatrade Groningen B.V. in Groningen, for which she had been sailing in the Seatrade Pool all along, took over the management. On 29 December 2006 the* ***Nostalgic*** *was sold for 6 million dollars to Green Reefers in Nesttun, Norway and her new name would be* ***Green Nostalgic****. On 11 June 2007 the reefer was renamed once again in* ***Nostalgic*** *and she starts sailing under Bahamian flag for B.V. Shipping Company Nostalgic in Nassau. She returns to the Seatrade Pool on 18 April 2007. Instead of the* ***Nostalgic****, which operationally speaking did not fit in the fleet of Green Reefers, the* ***Cap Gris Nez*** *of Seatrade (1990) went to Green Reefers for 9.3 million dollars and she was renamed* ***Green Chapeco****.*

Above: The ***Northern Express*** *loading seed-potatoes in the port of Beverwijk on 2 November 1989.*
Under: The superstructure of the ***Northern Express*** *moored along the Noorderkade in the port De Pijp of Beverwijk to load seed-potatoes on 22 December 1999.*

*The **Lindengracht** was launched at Miho Zosensho K.K. in Shimizu on 24 August 1988 as **Newca**, however, on 29 October 1988 she was delivered as **Kapitan Silin** to Far East Shipping Co. (Fesco) in Vladivostok. In February 1992 she would fly the Dutch flag as **Lindengracht** of C.V. Scheepvaartonderneming 'Lindengracht' in Amsterdam and managed by Spliethoff's Bevrachtingskantoor B.V. Her length is 113.03 metres and her breadth 19.23 metres. The ship has a gross tonnage of 6,037 and a deadweight capacity of 9,682 tons with room for 564 TEU. On deck are two deck cranes that can lift 50 tons each. In June 2000 the **Lindengracht** was sold and on 27 June 2000 in Valleyfield in Canada she was delivered to Umialarik Transportation in Montreal (Nunavut Eastern Arctic Shipping) and renamed **Umiavut**. She is managed by Transport Nanuk Inc. During the summer the ship is used to supply Inuit (Eskimo's). In the winter Spliethoff charters her as the **Umiavut** back again and is Amsterdam put on her stern again.*

*Above: The **Lindengracht** leaves IJmuiden for open sea on 31 March 1999.*
*Below: **Umiavut** on the North Sea Canal on 5 April 2002.*

The Nineties

After having been christened by Mrs. L.v.d. Poel-Splunter, the tanker ***Theodora****, built for the purpose of carrying bitumen, oil and chemicals, was launched at Scheepswerf en Machinefabriek De Merwede, v/h Van Vliet & Co, in Hardinxveld-Giessendam on 9 March 1991. Trials were carried out on the 5th of April 1991 and after that she was delivered to Scheepvaartmaatschappij Castor B.V. on the 12th of April 1991. Theodora Tankers B.V. in Rotterdam would be dealing with her management.* ***Theodora****'s length is 110.60 metres and her breadth 17.00 metres, with a Gt of 4,098 and a deadweight capacity of 6,616 tons. The tanker has nine cargo tanks with a total of 5.138m*^{3}*, which can be heated up to 250 degrees Celcius.*
In 1992 Theodora Tankers becomes a part of Paktank International (Pakhoed), which changes into Paktank Shipping (Pakhoed) in 1994. In 1999 the ***Theodora*** *is sublet to Bulgersteijn B.V. and is managed by Theodora Tankers B.V. in Dordrecht. In 2000 the tanker's management changes to Vopak Theodora Tanker Services B.V. in Dordrecht. On 14 april 2001 Theodora Tanker Services B.V. is taken over by the Swedish Tarbit Shipping AB in Skärhamn. The ship's name remains the same, just as her homeport Rotterdam. Her management remains dealt with by Theodora Tanker Services B.V. However, the grey colour of the hull is now replaced with a blue colour. The tanker mainly sails to north-western Europe and the Mediterranean.*

Above. The ***Theodora****, still with a grey hull*
Below: The coaster in the colours of Tarbit sailing on the Westerscheldt passing Flushing on 10 October 2006.

W&B
HELENE
DELFZIJL

On 24 May 1991, the **Helene***, having yard number 802, was launched at Niestern Sander B.V. in Delfzijl. The ship was christened by Mrs. I. Moilanen, spouse of A. Moilanen, managing director of Oy Finnterminals AB in Finland. The technical trails on the North Sea were on 10 July 1991 and the delivery to C.V. m.s. Helene in Delfzijl took place on 23 July 1991. The coaster is managed by Wijnne & Barends in Delfzijl. She is the second ship in a series of seven almost identical sisterships of the HC 3000 class. The* **Helene** *is 87.96 metres long and 12.58 metres wide. The gross tonnage amounts to 2,561 and the deadweight capacity to 3,284 tons. The* **Helene** *can transport 4,200 m^3 of timber in total. Room is available for 102 TEU in the cargo hold and 27 TEU on deck. On deck too stood a Radial side loading travelling crane of 15 tons, which has been removed at a later date. The engine room houses a 6-cylinder 4S. C.S.A. Caterpillar 3606TA diesel engine of 2,508 bhp, enough for a speed of 12,5 knots. The coaster is equipped with a reinforced bow and therefore she has an Ice Class 1A certificate meeting the requirements as drawn up in accordance with Finnish and Swedish Ice Class Rules dating from 1985. Her sailing radius is mainly the Baltic.*

Left: The **Helene** *on 14 April 2002, just after leaving the Zuidersluis (lock) in IJmuiden.*
Right: The coaster enters IJmuiden on 11 October 2003. The side loading travelling crane, which was removed later on, is clearly visible.

A real beauty of a ship is the **Oceaan Klipper**. *The reefer, with yard number 254, was launched at Yssel-Vliet Combinatie B.V. in Capelle aan den IJssel on the 17th of December 1991. Trials followed on 19 Februari and the delivery on 21 February 1992. The 108.22 metres long and 16.21 metres wide ship was owned by Jaczon Oceaan Klipper B.V. in Scheveningen and Jaczon Rederij & Haringhandel in Scheveningen would manage her. The Gt is 4,091 and the deadweight capacity 5,416 tons. The ship has a 6-cylinder 4S. C.S.A. Krupp MaK 6M552C diesel engine of 5,712 bhp, enough for 17 knots. The reefer has special side doors for lading bananas. Positioned on deck are four electrical deck cranes of 5 tons each. On the 7th of May 2006, the* **Oceaan Klipper** *was sold and officially delivered to B.V. Shipping Company Esmeralda in Larnaca, Cyprus. This a trading partnership within the Seatrade company, which also dealt with the management of this reefer. Incidentally, ever since her delivery the* **Oceaan Klipper** *already sailed in the Seatrade Pool. On 16 January 2007 she was sold to Green Shipping AS (Green Reefers), in Nesttun, Norway and renamed* **Green Ocean**.

Above: The **Oceaan Klipper** *moored along the Noorderkade in the port De Pijp of Beverwijk.*
Left & below: Just after leaving the Noordersluis (lock) in IJmuiden on 6 June 2000.

The **Katja** was launched at Scheepswerf Ferus Smit B.V. in Foxhol on 5 June 1993 and delivered after trials on 28 July 1993 to Shipping Company Ruby in Willemstad, Netherlands Antilles. Her management was dealt with by Sandfirden B.V. in Haren. The 89.56 metres long and 13.17 metres wide coaster had a gross tonnage of 2,735 and a deadweight capacity of 4,293 tons. On the 1st of January 1994 Rederij Katja B.V. became the new owner and she would now be managed by Hanno-OAM Shipping B.V. in Rotterdam (Heinrich Hanno & Co. N.V.). Her homeport now became Rotterdam too. North Sea Shipping B.V., a part of North Sea Holding B.V. (North Sea Petroleum), was participant in the coaster. In 1998 **Katja**'s management went to Arklow Shipping Nederland B.V. in Rotterdam. From 1994 until 1998 the coaster sailed from Europe to the east coast of America. As from 1998 until the present day the ship sails in north-western European waters. On 29 November 2005 the **Katja** was sold to Intermar Prosperity in Limassol and renamed **Helia**, sailing Cypriot flag. She is now managed by Craftchart in Tallinn, Estonia, however, the ship still sails in time-charter for Arklow.

Above: The **Katja** in the port De Pijp of Beverwijk, loading 4,000 tons of (artificial) fertilizer for Le Treport in north-western France on 5 November 2005.
Middle: On 16 September 1995 on the North Sea Canal just before entering side canal A in Velsen-Noord.
Below: The **Katja** on 23 December 2004, lying in the port of Beverwijk and with the colours of Arklow Shipping on her funnel.

MAVERIC

MAVERIC
GRONINGEN
Seatrade
SAETTA

*The **Maveric** was the last ship in a series of six almost identical refrigerated ships. She was launched at B.V. v/h Scheepswerven Gebr. van Diepen in Waterhuizen on the 27th of November 1992 and on 4 February 1993 she was delivered to C.V. Scheepvaartonderneming Maveric in Groningen. She was managed by Seatrade Groningen B.V. The **Maveric**, which can be classified as one of the most beautiful ships that ever sailed under Dutch flag, has a length of 135.70 metres and a breadth of 16.33 metres. Her Gt is 5,103 and her deadweight capacity is 6,106 tons. The total loading capacity amounts to 310,396 cbft. The **Maveric** has 8 side doors for the transport of bananas and next to this room is also available for 107 TEU. The reefers propulsion is provided by a 6-cylinder 4S. C.S.A. Krupp-MaK 6M1601C diesel engine of 8,973 bhp. On 27 December 2006 the **Maveric** was sold to Green Shipping AS (Green Reefers), in Nesttun, Norway and she was renamed **Green Maveric**.*

*Left page: The **Maveric** moored along the Kotterkade in the Haringhaven in IJmuiden on 19 April 2001.*
*Right: The **Maveric** moored along the Kotterkade (Cutter Quay) in the Haringhaven (Herring Port) in IJmuiden on 2 November 2002. Her lines are superfine.*
Below: Sailing into Southampton from the Solent in January 2007.

*The **Egelantiersgracht** and the **Egmondgracht** are almost identical sisterships. The **Egelantiersgracht** of C.V. Scheepvaartonderneming 'Egelantiersgracht' in Amsterdam was launched at Van der Giessen-De Noord in Krimpen aan den IJssel on 22 July 1994. Trials followed on the 29th and 30th of August and the delivery to the shipowner took place on 1 September 1994.*
*The **Egmondgracht** was launched at B.V. Merwede Shipyard in Hardinxveld-Giessendam on 16 July 1994. Her trials were on the 11th of October after which the delivery to the shipowner, C.V. Scheepvaartonderneming 'Egmondgracht' in Amsterdam, followed on 14 October 1994. Both ships are managed by Spliethoff B.V. in Amsterdam. Their length is 137.16 metres and their breadth 18.90 metres. The Gt is 8,447 and the deadweight capacity 12,754 tons. On deck three cranes can be found, all able to lift 60 tons. 730 TEU can be stored, 296 in the cargo hold and over 434 on deck. The total capacity in cbft is 556,900. Both ships have a 6-cylinder 4S. C.S.A. Wärtsilä Vaasa 6R46 diesel engine of 7,385 bhp.*

*Left: The **Egelantiersgracht** moored along the Buitenkade 3 (Outer Quay 3) in the Buitenhaven at Corus on 23 February 2003, loading yachts.*
*Below: On her way to sea we see the **Egmondgracht** on the North Sea Canal just in front of the Noordersluis (lock) in IJmuiden on 13 April 2002*

*The **Laurina** was launched on 22 August 1995 as **Laurina Neeltje** at Scheepswerf Peters B.V. in Kampen and on the 11th of November 1995 she was delivered to Rederij Gersom B.V. in Werkendam. She would be managed by Chr. Kornet & Zn. B.V. in Werkendam. The coaster was named after the mother of the 3rd generation of the Kornet family. The coaster has a Gt of 1,546 and a deadweight capacity of 2,335 tons. Room is available for 77 TEU. A 6-cylinder 4S. C.S.A. ABC 6 DZC diesel engine of 1,020 bhp provides the propulsion. On 18 April 2001 the **Laurina Neeltje** was sold to Zuidhinder B.V. This company was part of Damen Shipyards Gorinchem in Gorinchem. On 22 October 2001 the 85.14 metres long and 10.80 metres wide coaster was renamed **Laurina**. Her management remained at Chr. Kornet & Zn. B.V.*
As from 3 June 2002 the chartering was in the hands of Soetermeer, Fekkes Cargadoorskantoor B.V. in Zwijndrecht.
*On 21 March 2005 the coaster was sold to Stina Schiffahrts GmbH & Co. K.G., in Haren/Ems (Germany) and she was put into service, sailing under the Antigua & Barbudan flag, as **Stina**. The green hull now became blue. On the 24th of May 2006 captain Niels Bazuin becomes the new owner and he brings the coaster in in Noest Shipping B.V. in Roden. She is now renamed **Noest** and she is managed by Boomsma Shipping B.V. in Sneek.*

*Above: The **Laurina** enters IJmuiden on 5 October 2003.*
*Below: The **Noest** on 2 July 2007, just before entering the lock in Brunsbüttel.*

The keel of the ***Torne*** *was laid at Severnav SA Shipyard in Turnu, Severin in Romania on the 14th of March 1995. Due to the political climate the work was closed down and the ship would not be launched until 24 April 1996. She was put into service for Rederij Muonio B.V. in Papendrecht in May 1996. The coaster, named after a border river between Sweden and Finland, would be managed by Samshipping in Papendrecht. Samshipping was established by Mr. J. Berends, R. Teekman, A. Kuipers, A. Broekhuizen and G.H. Westerbroek in 1973. She had a length of 93.20 metres, a breadth of 15.00 metres, a Gt of 3,443 and a deadweight capacity of 4,812 tons. The coaster had room for 262 TEU. Propulsion was provided by a 6-cylinder 4S. C.S.A.Wärtsilä 6R32 diesel engine of 3,590 bhp. Primarily, the* ***Torne*** *carried cellulose, paper and timber mainly in North-European waters. In September 2004 the coaster was sold to Sandfrakt Rederi in Olensvaag in Norway. She now sails as* ***Norne*** *with Gibraltar as her 'flag state'.*

Above & below: Entrance of the ***Torne*** *in IJmuiden on 11 October 2001.*
Next page: On port side we see the ***Flinterdam*** *(1995) sailing toward open sea.*

TORNE
PAPENDRECHT

At Bodewes Scheepswerven B.V. in Hoogezand, the **Swallow** was launched on 24 April 1996, having yard number 571. After her trials on the 25th of May 1996 the ship was delivered to C.V. Scheepvaartonderneming 'Swallow 2' in Hoogezand. Several parties had interests in the ship, of which Bodewes had a majority. The coaster, with a length of 90.46 metres and a breadth of 13.20 metres, was managed by B&N Kustvaartbedrijf Moerman B.V. in Rotterdam, they moved to Ridderkerk later on, who also had a 16% share in the ship. At that moment Moerman was part of the Swedish Bylock & Nordsjöfrakt of Gothenburg. As from March 2005 Moerman operates as an independent company again. The Gt of the **Swallow** was 2,848 and the deadweight capacity 4,251 tons. In the **Swallow**'s engine room stood a 6-cylinder 4S. C.S.A. Stork-Wärtsilä 6SW280 diesel engine of 2,501 bhp. Room was available for 206 TEU. The **Swallow** mainly sailed European waters but she also sailed to America many times.
On 25 February 2003 the coaster was sold and in Delfzijl she was delivered to Briese Schiffahrts GmbH & Co. K.G., in Leer, who renamed her **Westerland**. Presently, she sails again as **Swallow** for Logistica Integral in Cee, Spain.

*Above: The **Swallow** moored along the quay of the 2nd Rijksbinnenhaven in Velsen-Noord on 25 January 2002.*
Middle: The coaster on her way to sea on the North Sea Canal on 31 May 2002.
*Below: The **Swallow** on the North Sea Canal on 21 March 2002.*

*The **Klostertal**, originally to be named **Otztal**, was launched at Severnaya Verf (Nordwerft) in St Petersburg as yard number 434 on 5 April 1996.*
She was completed at Neptun Industrie GmbH in Rostock. The ship was delivered in August 1996 to Geschwister Krohn Schiffarhrts GmbH & Co K.G. IX in Vienna and she is part of the Krohn & Co Gruppe while Alstership GmbH & Co. K.G. in Hamburg is her shipbroker. The somewhat angular ship is 108.75 metres long and 17.80 metres wide. The Gt amounts to 5,622 and the deadweight capacity to 7,000 tons and she has room for 342 TEU. A 6-cylinder 2S. C.S.A. B&W-Bryanski 6L35MC diesel engine of 4,568 bhp provides the propulsion.

*In December 2001, **Klostertal**'s flag becomes Dutch and her new owner is going to be Rederij K&T Holland C.V. in Delfzijl. On the 10th of December 2001 the **Klostertal**'s management transferred to Flagship Management Company B.V. in Farmsum.*

*Both pictures: The **Klostertal** on 7 May 2002, still in red colours, just after leaving the Noordersluis (lock) in IJmuiden. Later on the ship's hull would become blue.*

*The **Gaastborg**, number four in a series of seven almost identical ships, was launched at Bodewes Scheepswerf 'Volharding' Foxhol B.V. in Foxhol on 11 May 1996. After trials on the North Sea on the 18th and 19th of June, the ship was delivered to C.V. Scheepvaartonderneming 'Gaastborg' in Delfzijl. Her management came in the hands of Wagenborg Shipping B.V. in Delfzijl. The **Gaastborg**, being 90.20 metres long with a breadth of 13.70 metres, has a Gt of 2,820 and a deadweight capacity of 4,135 tons and she has room for 211 TEU. The engine room houses a 8-cylinder 4S. C.S.A. Stork-Wärtsilä 8SW280 diesel engine of 2,500 bhp. On 28 June 2002 Esmeralda Shipping N.V. in Willemstad becomes the new owner and the ship's flag becomes a flag of convenience: of the Netherlands Antilles. Esmeralda Shipping is a subsidiary of Astrid W. Waller K.G. in Willemstad. She is now managed by G. Waller in Hemmoor in Germany. However, her chartering is still in the hands of Wagenborg.*

Above: During trials on 19 june 1996 on the North Sea.
Below: Entering the Noordersluis (lock) in IJmuiden on 1 May 1999.
*Next page: The **Gaastborg** with Willemstad as her home-port, discharging a cargo of timber in Honfleur on 26 May 2005.*

IMO 9129122
GAASTBORG

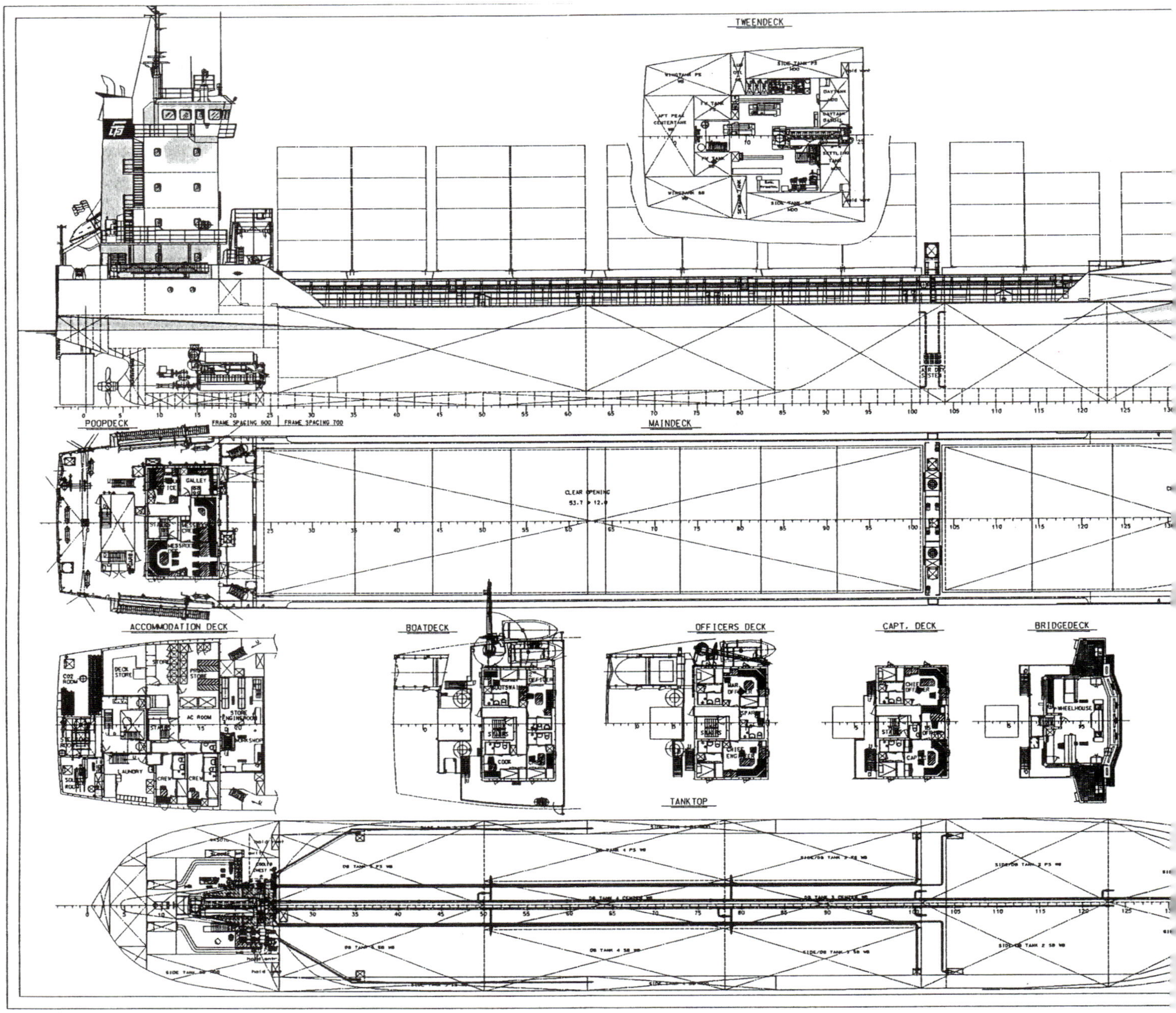
TWEENDECK
POOPDECK
FRAME SPACING 600
FRAME SPACING 700
MAINDECK
ACCOMMODATION DECK
BOATDECK
OFFICERS DECK
CAPT. DECK
BRIDGEDECK
TANKTOP

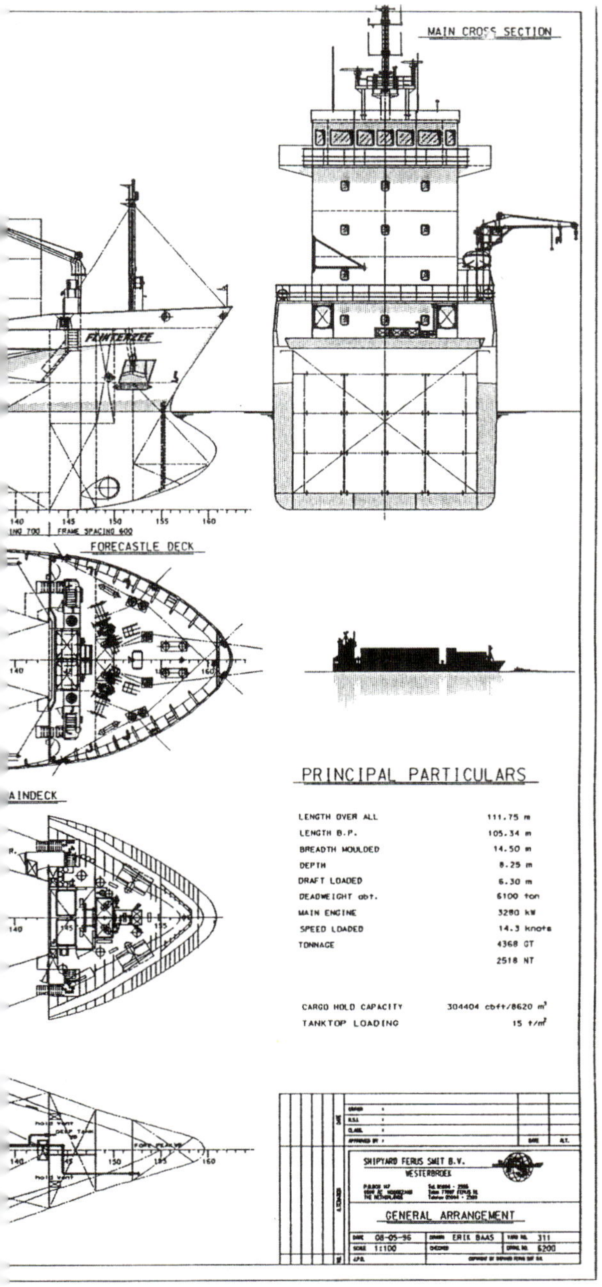

On the 10th of May 1997 the **Flinterzee** was launched athwartships at Scheepswerf Ferus Smit B.V. in Westerbroek. Following the official trial trip on the Ems, the **Flinterzee** was delivered to C.V. Scheepvaartonderneming Flinterzee in Groningen on 19 June 1997.
Her management was carried out by Flinter Groningen B.V. in Groningen. North Sea Shipping B.V., a part of North Sea Holding B.V. (North Sea Petroleum) is participant in the coaster. The 111.75 metres long and 14.60 metres wide ship was chartered by Wagenborg Shipping B.V. in Delfzijl. The **Flinterzee** has a Gt of 4,368 and a deadweight capacity of 6,150 tons. A 8-cylinder 4S. C.S.A. Stork-Wärtsilä 8R32E diesel engine of 4,460 bhp provides propulsion with a maximum speed of 14.5 knots. As from mid June 1999 the **Flinterzee** was chartered by Ancora AFS in Rotterdam. In July 2006 the ship was sold to foreign interests (Bergen Banking Shipping AS in Bergen) but she continued to sail with her own name and Dutch flag for Flinter, now on bare-boat basis. On 15 September 2006 the **Flinterzee** joined Flinter Bareboat B.V. in Groningen and as from 29 September 2006 Flinter Shipping B.V. in Rotterdam is her shipbroker.

All pictures: The **Flinterzee** sailing towards open sea on the Westerscheldt near Flushing on 26 September 2002.

*The **Isabel** enters Archangelsk on 10 April 2004. On port side we see the Russian river-icebreaker **Kapitan Chadaev**. The ship was built in 1978 and is owned by Volga River Shipping Co. The length is 77.60 metres and she is 16.30 metres wide. A diesel-electric engine of 4,490 bhp provides the propulsion.*

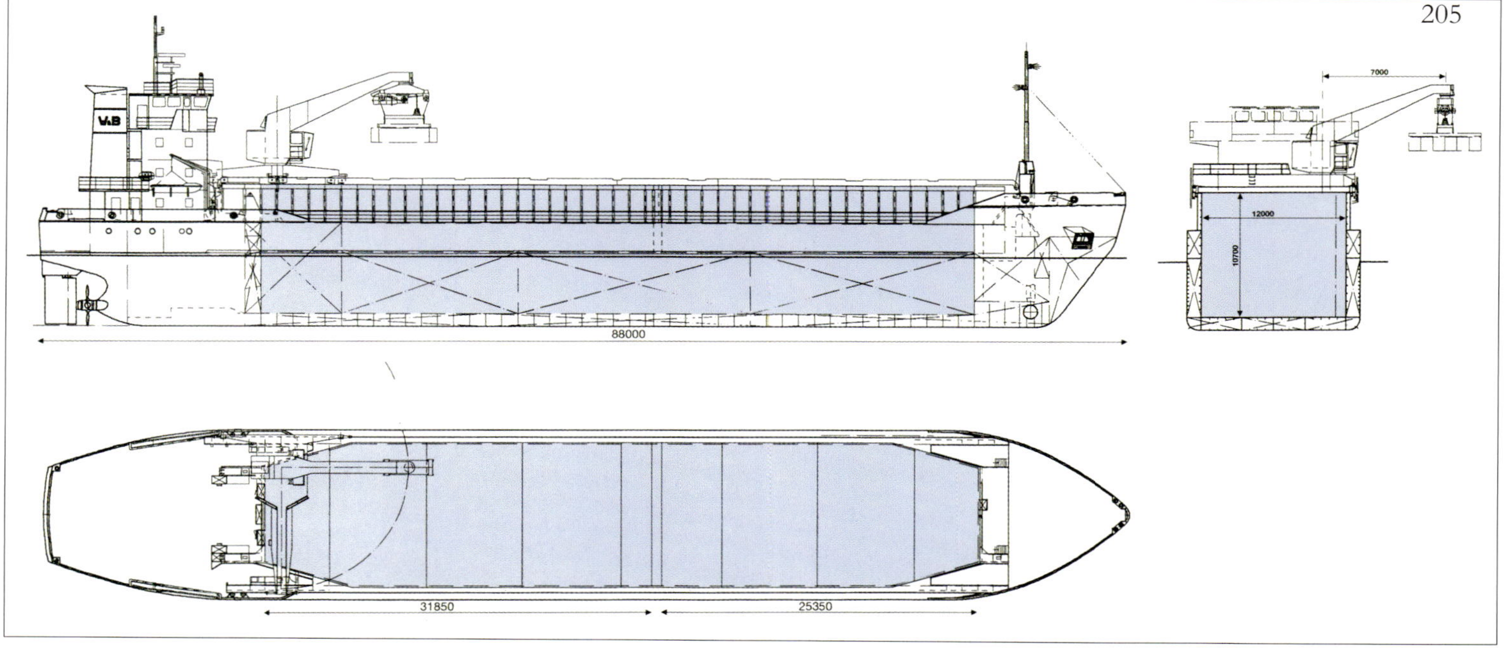

At Niestern Sander B.V. in Delfzijl the **Isabel** was launched on 8 November 1997. The trial trip on the Ems took place on 18 December and on the 22nd of December 1997 she was delivered to C.V. Scheepvaartonderneming Isabel in Delfzijl. She would be managed by Wijnne & Barends in Delfzijl.
The **Isabel** was the last ship in a series of three ships of the 900 Fast Handling Flexibox Carrier type. She has a length of 88.05 metres and a breadth of 14.46 metres, with a Gt of 3,323 and a deadweight capacity of 4,182 tons and is able to transport a total of about 6,200 m³ of timber. The engine room houses a 8-cylinder 4S. C.S.A. Caterpillar diesel engine of 3,346 tons. A 20 tons parallel side loader crane was positioned on deck, however, that crane was removed at a later date. The **Isabel** is equipped with a reinforced bow (Ice Class 1A) to enable her to sail through ice fields. In September 2003, C.V. Scheepvaartonderneming Isabel II became the new owner. At present the coaster sails for B.V. Beheersmaatschappij Isabel II in Delfzijl. This company is owned by captain-owner Willem Hoogenhout for 39%, Spliethoff B.V. for 51% and W&B for 10%. The **Isabel** runs a regular service from Archangelsk via the Arctic Cape to Beverwijk carrying sawed timber. At the time this book was published the timber coaster was completing her 65th tour, which on average lasts 22 days.

*Above: The **Isabel** moored along her permanent discharge quay (Noorderkade/Northern Quay) in the port De Pijp of Beverwijk on 7 May 2002, still equipped with a parallel side loader (travelling) crane.*
*Right: Entrance of the **Isabel** in Archangelsk on 5 April 2005.*

*The keel of the **Thalassa**, with yard number 454, was laid at Scheepwerf Peters B.V. in Kampen in April 1998. The launch of the 89.25 metres long and 13.40 metres wide coaster was in December 1998. Trials of this type Polar Combi Porter 200 coaster on the North Sea were carried out on the 5th of February 1999 after which the ship was delivered to Geert Switynk in Harlingen. The Gt of the **Thalassa** is 2,780 and her deadweight capacity is 3,791 tons, with room for 200 TEU. The ship has a 6-cylinder 4S. C.S.A. MaK 6M25 diesel engine of 2,448 bhp. On 16 January 2003 Thalassa B.V. in Harlingen became the owner. On 20 January 2006 the **Thalassa** was sold to Scheepvaartbedrijf A.J. Kok in Maartensdijk and she was renamed **Liamare**. Liamare was derived from the first names of Adriaan Kok's wife and children. The **Liamare** was managed by Flinter Shipping B.V. in Rotterdam. On 21 December 2006 the **Liamare** was already sold again. She is now in service as **Nordgard** of Bore Nordgard B.V. and she is managed by Bore Shipowners B.V. in Rotterdam. The ship mainly sails in north-western Europe.*

*Above: The **Thalassa** discharges a load of timber at the Noorderkade in Beverwijk on 8 March 2001. Clearly visible are the fine lines of the wheel house, providing a good view on all sides.*
*Left: The **Thalassa** in ballast outward bound to sea on 26 September 2001, just after leaving the Middensluis (lock) in IJmuiden.*

The **Tramper** belongs to the (Confidence) multi-purpose container ship type, but she could just as easily be categorised as heavy-cargo ship. The keel of the **Tramper** was laid on 21 November 1998, at Zhonghua Shipyard in Shanghai and her launch was on the 4th of February 1999.
The delivery to West African Shipping N.V. in Willemstad was on the 18th of May 1999. Her managment was in the hands of Mammoet Shipping B.V. (70% Spliethoff) in Amsterdam. Subsequently, the **Tramper** left Masan for Wismar with a number of marine engines as her cargo. Next she headed for Schiedam, where she arrived on 26 July. Here, at Huisman-Itrec two enormous deck cranes were installed, each of them able to lift 275 tons. Spliethoff B.V. in Amsterdam became the full owner of Mammoet Shipping on 21 November 2000. As of 1 January 2001 this name was changed into BigLift Shipping B.V. On 25 March 2004 C.V. Scheepvaartonderneming Tramper became the owner of the ship. In Rosario, on the 6th of April 2004, the **Tramper**'s flag changed from Antillean into Dutch flag while her management remained at BigLift Shipping B.V. in Amsterdam. The **Tramper**, the first ship in a series of four Tra-type ships is 100.70 metres long and her breadth is 20.40 metres, with a Gt van 6,583 and a deadweight capacity of 8,874. The engine room houses a 8-cylinder 4S. C.S.A. Wärtsilä NSD 8L46B diesel engine of 10,608 bhp.
The Tramper can store 195 TEU in her cargo hold and 176 TEU on deck.

Above: The **Tramper** in the Welland Canal in Canada on 4 August 2003.
Below: At Bilboa on 13 March 2000.

*Originally, the **Bosgat** would be built at Santierul Naval Mangalia SA in Mangalia in Romania. However, this was cancelled and instead the ship was built at Daewoo Mangalia Heavy Industries Co. in Mangalia, where she was launched on 18 March 1999. Mid November 1999 the 87.96 metres long and 12.39 metres wide coaster was deliverd to C.V. Scheepvaartonderneming Bosgat in Harlingen, of which Mr. Wessel L. Feenstra was the single owner. The ship has a Gt of 2,435 and a deadweight capacity van 3,697 tons. Propulsion is provided by a 8-cylinder 4S. C.S.A. Wärtsilä NSD 8FDHD240 diesel engine of 2,000 bhp. On the 4th of July 2003 Jan and Wim van Veen of Zwartsluis became the new owners of the coaster and they renamed her **Ramspol**. She was brought in in Ramspol B.V. The brothers sailed the coaster until 10 October 2006, when Rana Shipping C.V. in Harlingen became the new owner and the name of **Rana** appeared on her bow. This C.V. is run by the original owner Wessel Feenstra, his brother Ysbrugt, E.P. Shipping & Trading B.V. from Rozenburg and Mr. Peters. Each of them owns 1/3 of the ship, except Mr. Peters who participates for 10%. The **Rana**'s management is carried out by Flagship Management Company B.V. in Delfzijl, who deals with all crew matters, amongst others. E.P. Shipping is the shipbroker. The ship mainly sails to north-western Europe.*

*The **Bosgat** on her way to sea, just after leaving the Middensluis (lock) in IJmuiden on 15 June 2002.*

A new century

On the 24h of February 2001, the ***Polar Sky*** *was launched at Scheepswerf Peters B.V. in Kampen. In order to complete the trials on the North Sea on the 20th of March the coaster of the Polar Combi Porter 200 type had to sail via the IJsselmeer (IJssel Lake) and through the lock at Kornwerderzand. On 23 March 2001 she was delivered to C.V. Scheepvaart-onderneming Polar Sky in Delfzijl while RAH Management B.V. in Rhoon took care of her management. Her shipbroker would be Royal Shipping B.V. Both companies were part of the Royal-Amasus Hansen Group. As from 1 October 2001 Amasus Chartering B.V. in Delfzijl was the shipbroker. However, in 2003 all companies faced financial problems and they were forced to close down their activities. At the end of August 2003 the management of the* ***Polar Sky*** *was taken over by Q-Shipping B.V. in Rhoon. The* ***Polar Sky*** *is 89.25 metres long and with a breadth of 13.40 metres, a Gt of 2,863 and a deadweight capacity of 3,792 tons, she has room for 208 TEU. In 2004 the ship was sold to Briese Schiffahrts GmbH & Co. K.G. in Leer and renamed* ***Frisian Sky*** *while now sailing under Gibraltar flag.*

Aboven: The ***Polar Sky*** *when leaving the Noordersluis (lock) in IJmuiden and below she is seen on the North Sea Canal on 9 September 2002.*

At Detlef Hegemann Rolandwerft GmbH & Co. K.G. in Berne in Germany the container feeder ***Merwedijk*** *was launched on 17 February 2001. The ship was christened by Mrs. Mol, spouse of one of the managing directors of De Vereenigde Compagnieeën B.V. (shipping investments). The trials at the North Sea took place on the 12th of April and the delivery to C.V. Scheepvaartonderneming ms Merwedijk in Groningen (Mare Trader B.V.) followed on the 14th of April 2001. The* ***Merwedijk*** *belongs to the group of Mare Shipmanagement B.V., which in itself is part of Navigia Shipping B.V. in Groningen. The entire commercial, administrative, nautical and technical management is carried out by Navigia Shipmanagement B.V. The ship's length is 132.23 metres and her breadth 19.40 metres. The Gt is 6,368 and the deadweight capacity 8,298 tons. The* ***Merwedijk*** *has a 8-cylinder 4S. C.S.A. MaK Caterpillar 8M43 diesel engine of 9,788 bhp, enough for a speed of 17 knots. She has room for 707 TEU.*

From the start the ship has been put in service for the Danish Unifeeder A/S of Aarhus. Her sailing radius mainly covers north-western Europe and Hamburg, Bremerhafen and the Baltic in particular with an occasional visit to Rotterdam and Antwerp.

Both pictures: The ***Merwedijk*** *sailing on the Westerscheldt near Flushing, on her way to Antwerp on 12 October 2006.*

The keel of the ***Montana*** *was laid, as yard number 201, at ATVT Sudnobudivniy Zavod 'Zaliv' in Kerch in the Ukraine on 1 September 2000. Only the ship's hull would be built at this shipyard and the launch was on 6 February 2001. On 19 February 2001, the tug/supplier* ***Thomas de Gauwdief*** *(1971) started to tow the ship to Rotterdam, where it would arrive on 30 March 2001. At Machinefabriek Olthof B.V. in Capelle aan den IJssel the engine was installed and the ship completed. IHDA Shipbuilding Service B.V. in Krimpen aan den Lek commissioned the overall building operation. Trials were completed on 25 July 2001 and the coaster was put in service on the 4th of August 2001. Owner of the* ***Montana*** *was Rederij Montalex from Vogelwaarde. Messrs. Pieter Pomp from Vogelwaarde, Willem Oose and Gerrit Visser were the men behind this shipowner. As from 1 January 2004 Pieter Pomp became the single owner of the ship. The* ***Montana*** *is named after the daughter of Pieter Pomp and the name Montalex is an abbreviation of the names of the daughter (Montana) and son (Alexander) of Pieter Pomp. The* ***Montana*** *was 107.45 metres long and 15.32 metres wide and she had a Gt of 4,112 and a deadweight capacity of 5,826 tons. A 6-cylinder 4S. C.S.A. MaK Caterpillar 6M32 diesel engine of 3,915 bhp provided a speed of 13,5 knots. She had room for 325 TEU. The* ***Montana*** *mainly sailed in Northern European waters. On 27 April 2005 the* ***Montana*** *was sold to Aasen Transport in Mosterhamn in Norway. She was renamed* ***Aasheim****.*

The ***Montana*** *moored along the Noorderkade in the port De Pijp of Beverwijk on 16 September 2003. The ship just discharged a cargo of wood chips.*

Left: Operating on icy waters on 12 April 2003.
Below: Bow waves spraying the deck on 17 January 2003.

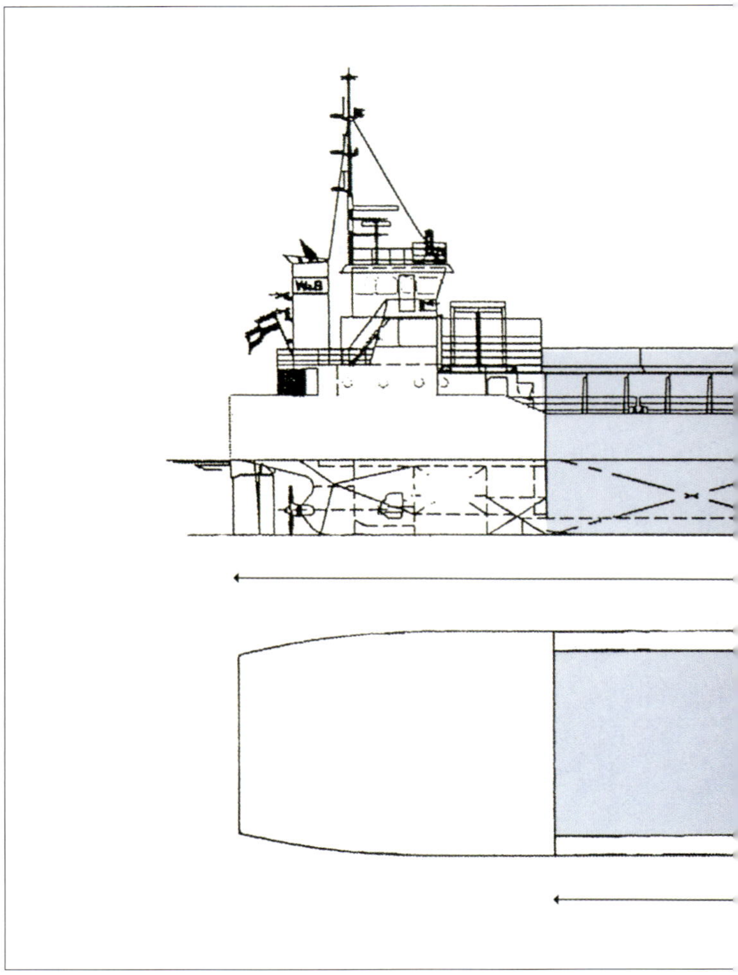

*The **Nora** and **Nola** are of the VHC 2000 (Very High Cubic) type and they were specially designed with spherical hatches in order to be able to carry more cargo below deck. Both ships were built at Koninklijke Niestern Sander B.V. in Delfzijl. The **Nora** was launched on 7 September 2001, the technical trial trip took place on the 18th of October 2001 and she was delivered on 1 November 2001. The launch of the **Nola** was on 11 January 2002 and she was delivered on the 1st of February 2002. Both ships sail in service for Scheepvaartonderneming Opwierde B.V. (Timberfleet C.V.). Timberfleet is a subsidiary of Wijnne & Barends B.V. in Delfzijl. The latter also manages both ships. These almost identical ships (they differ in their radar masts amongst others) have a length of 79.99 metres and a breadth of 12.10 metres, with a Gt of 1,978 and a deadweight capacity of 2,050 tons. The cargo capacity for timber below deck is 3,000 m³. A 12-cylinder Caterpillar 4S. C.S.A. 3512B diesel engine of 1,700 bhp provides the propulsion.*
The ships mainly transport timber, paper, and bulk cargo from Scandinavia and Russia to Western European ports.

*Left page, above: The **Nora** seen here just before entering the Middensluis (lock) in IJmuiden on 25 July 2004, was christened by Mrs. N.M. Walsarie Wolf-Cox. She is the great- great-grand-daughter of co-founder Barend Cornelis Barends.*
*Right: The **Nola** seen here moored along the Noorderkade of port De Pijp in Beverwijk on 23 June 2002, was christened by Mrs. B.M.V. Brakel-Uiterwaal. This lady too is the great- great-grand-daughter of the co-founder of the company.*

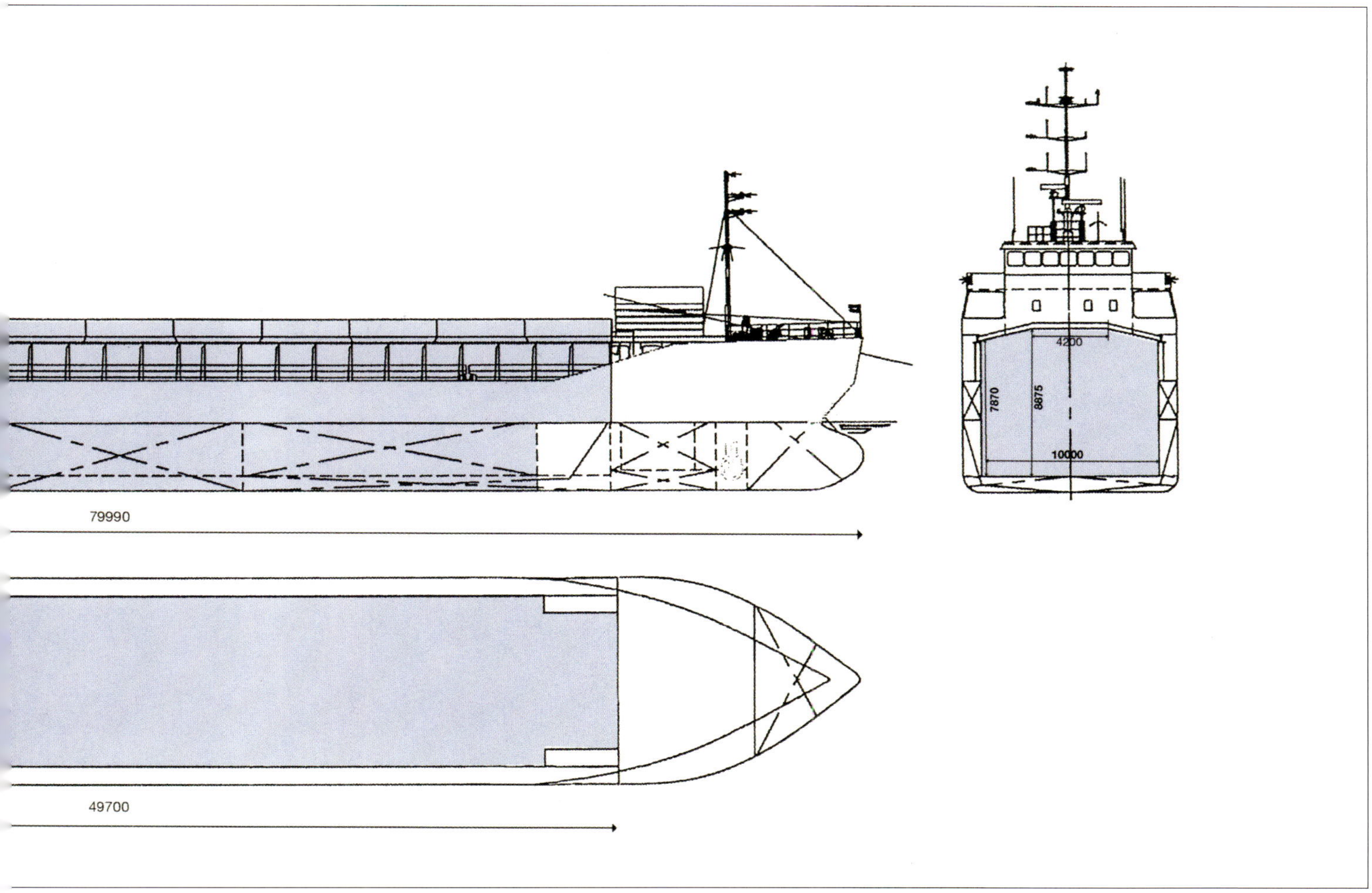

*The **Daniel K**, having yard number 1006, was launched at Rousse Shipyard JSC in Rousse in Bulgaria. The delivery to Rederij K&T Holland C.V. IV in Delfzijl was on the 24th of October 2002. On 22 November 2002 the **Daniel K** was christened in the Netherlands and on the 29th of December 2002 her flag officially became Dutch. The **Daniel K** is managed by Flagship Management Company B.V. in Farmsum and her chartering is dealt with by Altership GmbH & Co. K.G. in Hamburg.*
The ship is 90.25 metres long and 15.20 metres wide. She has a Gt of 3,037 and a deadweight capacity of 4,247 tons. A 8-cylinder 4S. C.S.A. MaK Caterpillar 8M25 diesel engine of 3,218 bhp provides the propulsion. Room is available for 325 TEU. The ship mainly sails in Northern European waters.

*The **Daniel K** leaving the Middensluis (lock) in IJmuiden on 9 March 2005.*

*The chemicals/oil tanker **Bro Gemini** of the Bro G type was launched athwartships at Scheepswerf Ferus Smit B.V. in Westerbroek on 8 February 2003. The ship was christened by Mrs. Betty van den Driest, wife of the managing director of Vopak. The delivery to C.V. Scheepvaartonderneming Bro Gemini in Delfzijl took place on the 4th of April 2003. She is owned by Eric Thun AB in Lidköping. The management is carried out by Marin Ship Management B.V. in Farmsum while the ship sails in service for Broström Tankers AB in Gothenburg. The tanker has a length of 114.95 metres and a breadth of 15.30 metres, with a Gt of 4,107 and a deadweight capacity of 7,178 tons. The tanker's engine room houses a 4 t.e.w. 6-cylinder 4S. C.S.A. Wärtsilä 6R32E diesel engine of 3,300 bhp.*

*Both pictures: The **Bro Gemini** entering IJmuiden on 12 April 2004.*

On the 19th July 2003, the **Onego Trader** was launched as **Velserdiep** at JSC Zaliv Shipyard in Kerch in the Ukraine. After that the hull was towed to the Netherlands, where it arrived in Delfzijl in September. The ship was completed at Bodewes Scheepswerven B.V. in Hoogezand. The coaster was christened as **Onego Trader** by Mrs. Olga Kremeskaia, wife of Mr. Kremenskiy, managing director of Onego Shipping Ltd., SPb, in St Petersburg. On the 18th of December 2003 she was delivered to C.V. Scheepvaartonderneming ms Velserdiep in Groningen. She is managed by Feederlines B.V. in Groningen. Straight after her delivery the **Onego Trader** would start to sail in charter for Onego Shipping & Chartering B.V. in Rhoon. The coaster has a length of 119.98 metres and a breadth of 15.20 metres. Her Gt is 5,052 and the deadweight capacity is 7,650 tons. The coaster of the box singledeck type has two deck cranes of 40 tons each. Propulsion is provided by a 8-cylinder 4S. C.S.A. MaK Caterpillar 8M32 diesel engine of 5,222 bhp. The ship sails on all the world's seas.

*Left: The **Onego Trader** returns in Delfzijl after her second trial trip on 19 December 2003. The next day she would set off for her first voyage.*
*Above and below: The **Onego Trader** moored in the IJmondhaven in IJmuiden for bunkering on 13 August 2005.*

*The hull of the **Diamant** was built at Shipyard Leda D.O.O. in Korcula in Croatia and this was launched on 28 May 2005. It was completed at Scheepswerf Peters B.V. in Kampen. Trials took place on the Ems on the 17th of October 2005 and the delivery to C.V. Scheepvaartonderneming Diamant (Van Donderen Combinatie) in Delfzijl followed on 21 October. This C.V. is run by Messrs. Klaas van Donderen, Arthur Schwerink and Jamie Muylaert. The ship was given her name by the wife of Klaas van Donderen, who sensed the meaning of power and beauty in diamant.*

*The **Diamant** is of the Super Saimax type and she is completely boxed shaped. The so-called short-sea-container carrier has been reinforced to enable her to sail through ice fields and she is fit for sailing through the Finnish Saimaa Lakes complex. Her length is 82.50 metres and her breadth 12.50 metres, with a Gt of 2,409 and a deadweight capacity of 3,217 tons. Her deadweight tonnage when sailing on the Saimaa Lakes is 2,600 tons. A 12-cylinder 4S. C.S.A. Caterpillar 3512BHD DITA diesel engine of 1,700 bhp is sufficient for a speed of 12 knots. The coaster has room for 132 TEU, 36 of which on the hatches. She is registered for Ice Class 1C. Wagenborg Shipping B.V. is her shipbroker. After the delivery, the **Diamant**, commanded by captain/owner Klaas van Donderen, sailed to Harlingen to load salt for Joutseno in Finland.*

*The **Diamant** sailing on the Westerscheldt near Flushing on her way to open sea on 10 October 2006.*

TERSCHELLING

*The **Terschelling**, having yard number 362, was launched at Scheepswerf Ferus Smit B.V. in Westerbroek on 4 November 2005. Mrs. Berber Switynk-Ishtia christened her. The ship's trials took her to the Ems on the 5th and 6th of December 2005. The delivery to Terschelling Shipping C.V. in Harlingen was on the 9th of December 2005. The **Terschelling** is managed by Kustvaart Harlingen B.V. in Harlingen. This company is led by four captains/owners, i.e. Robin Switynk, Roelof Rientjes, Johan Oomens and Arno Boere. The ship is 110.78 metres long and 14.00 metres wide. Her Gt amounts to 3,990 and the deadweight capacity to 6,080 tons. The engine room houses a 8-cylinder 4S. C.S.A. Mak Caterpillar 8M25 diesel engine of 3,585 bhp, providing the coaster with a speed of 14 knots. The ship has a Ice Class 1A registration and her sailing radius will mainly be around Europe. After the delivery and commanded by captain Johan Oomens, she left Harlingen for her first voyage to Karlsborg and Sundsvall, where she will load cellulose for Pasajes.*

*Left page: Launch of the **Terschelling** at Scheepswerf Ferus Smit on 4 November 2005.*
*Above and below: Entering **IJmuiden** on 23 March 2007.*

*Next page: The **Isabel** moored along the Noorderkade on 6 November 2005 in the port De Pijp of Beverwijk. Behind her we see the **Scot Mariner** and **Royal Cooler**.*

ISABEL

List of ships'names

Sources consulted

J.H. Anderiesse, E.A. Kruidhof, J. Oostmeijer:
Carebeka 1939-1983. World Ship Society, Kendal 1995.

Jan Anderiesse & Leslie Spurling:
Gruno 1937-2002. Uitgeverij De Alk B.V., Alkmaar 2003.

Hein Bekenkamp:
Sandfirden Worldwide 1984-1989. Wij varen op volle kracht vooruit. Sandfirden, Haren 1989.

Hans Beukema:
Wagenborg 100 jaar, Transport over water, wadden en wegen. Uitgeverij De Alk B.V., Alkmaar 1998.

K.W.L. Bezemer:
Geschiedenis van de Nederlandse Koopvaardij in de Tweede Wereldoorlog-deel 1. Elsevier, Amsterdam/Brussel 1986.

Gerrit de Boer e.a.:
Wagenborg 100 jaar, De schepen van 1898 tot 1998. Uitgeverij De Alk B.V., Alkmaar 1998.

G.J. de Boer:
150 jaar van Uden. Uitgeverij De Alk B.V., Alkmaar 1998.

G.J. de Boer:
Scheepvaart-jaarboek. Diverse jaren. Uitgeverij De Alk B.V., Alkmaar.

A. Boerma:
Coasters. Schepen van de kustvaart toen en nu. Uitgeverij De Alk B.V., Alkmaar 1985.

A.C. van Dorp e.a.:
Varen & Opslaan. 75 jaar Gebr. Broere, 50 jaar Rederij Theodora. Uitgeverij De Alk B.V., Alkmaar 1997.

Drs. D. Gorter - G.J. de Boer:
Standaardschepen 1939-1945 in de Nederlandse en Belgische koopvaardij, 2 delen. Uitgeverij De Alk, Alkmaar 2005.

F. Haalmeijer, B. Scholten:
Van Nievelt, Goudriaan & Co's Stoomvaart Maatschappij (NIGOCO). Uitgeverij De Alk B.V., Alkmaar 1995.

P. van Heck:
Vlootparade, Eigen beheer, Dordrecht 2000.

Wilco van Koldam, Harm van der Veen en Jan Nico Wilkens:
Veenkoloniale Zeevaart. Schuur Veendam B.V., Veendam 1979.

Lloyd's London:
Lloyd's Confidential Index of Steam and Motor Vessels. Diverse jaren. Lloyd's of London Press Ltd, Londen.

Lloyd's Register:
Lloyd's Register of Shipping. Diverse jaren. Lloyd's Register, Londen.

Willem H. Moojen:
Nederlandse Koopvaardij in beeld 1950-1959 (I). Lanasta, Emmen 2006.

Willem H. Moojen:
Nederlandse Koopvaardij in beeld 1960-1969 (I). Lanasta, Emmen 2006.

Willem H. Moojen:
Hollandsche Stoomboot Maatschappij. De vaart op het Verenigd Koninkrijk en West-Afrika. Deel 2: De vloot. Lanasta, Emmen 2006.

Moorman:
Moorman's Jaarboek voor Scheepvaart en Scheepsbouw. 1930-1971. Moorman's Periodieke Pers, Den Haag.

Frederik Muller:
Schippers-Slepers-Reders. 225 jaar scheepvaartgeschiedenis van de familie Muller. Scheldt Information Service, Terneuzen 2006.

L.L. von Münching:
De Nederlandse Koopvaardijvloot in de Tweede Wereldoorlog-deel 2. Den Boer Uitgevers, Middelburg 1986.

L.L. von Münching:
De geschiedenis van de Batavier Lijn. Uitgeverij Van Wijnen, Franeker 1994.

Drs. F. Post:
Groninger Scheepvaart en Scheepsbouw vanaf 1600. Profiel Uitgeverij, Bedum 1997.

Bob van Raad, Teun van der Zee en Roel Zwama:
Thuishaven Rotterdam. Scheepsfotoruilbeurs, Rotterdam 2001.

Seehafen Verlag:
The German Merchant Fleet. Die Deutsche Handelsflotte 1982-1983. Seehafen Verlag Erik Blumenfeld GmbH & Co. K.G., Hamburg 1982.

Roy Stemman:
Geest 1935-1985. Geest Industries Ltd./EP Consultants, Stamford 1986.

Uitspraak Raad voor de Scheepvaart 2001 nr 18.

Photographs

Foto Aerophoto Eelde: 200b
Foto J. Akkerman: 21o, 129o
Foto Ambuscade Marine Photography: 157o
Foto Ron Baker: 118b
Foto Piet Boonstra: 91b, 108lb
Foto G. Bunschoten: 15, 16o, 24, 32b, 33b, 36o, 39m, 41, 49b, 51ro, 56o, 57b, 58o, 62o, 81b, 83o, 88o, 93o, 95b, 100b, 102b, 103b, 144m
Foto J.K. Byass: 133b
Foto John G. Callis: 19b, 27o
Foto Jeff Cameron, Willem H. Moojen: 207b
Foto John & Marion Clarkson, Longton: 25o, 57o, 71b, 73o, 77o, 85m, 89o, 103m, 106m, 109b, 119b, 120o, 141m
Foto F.H. van Dijk: 111b
Foto Dijkstra: 32o, 42b, 45o, 48o, 65b, 67b, 76b, 79o, 88b, 90b, 94b, 104b, 106b, 107b, 109o, 112b, 116o, 123b, 125b
Foto David Dixon: 147o
Foto C.M. Drayson: 193o
Foto A. Duncan: 63o, 78b
Foto van Eldik: 150b
Foto Feederlines B.V.: 216m
Foto Folkers: 67o
Foto Fotoflite, Ashford: 17, 18b, 28, 80, 82o, 105o, 115b, 129m, 140b, 141o, 159b, 187b
Foto Peter Glenn: 118o
Collectie D.J. Goubert: 160b
Foto C.A. Hill: 33o
Foto Dave Hocquard: 41o, 46o, 73m
Foto E.A. Hof: 38b
Foto G.J. Hosmar: 55
Foto L.M. Johannes: 164b
Foto J. Klaver: 61b
Foto C. Kramer: 110
Foto Tom Kroeze: 72
Foto La Cigogne: 126b
Foto Leeuwarder Courant: 101
Foto M. Lindenborn: 44o, 53b, 57m, 111o, 130b, 133o
Foto Donald Macfie: 37o, 69o, 74b, 79b
Foto Foto-Mar: 103o
Foto F. Meijer: 75b
Foto Marco Middel, Wijnne & Barends: 204, 205o, 212m, o
Foto G. Montenij: 18o, 27b, 45b, 63b, 64o, 84o, 85b, 104m, o, 115o, 123o, 126o, 127
Foto Willem H. Moojen: 6, 12, 39b, 40b, 51lo, 54, 66, 70o, 74o, 77b, 81o, 84b, 87b, 90o, 94o, 97m, 98b, 99, 116b, 121, 122o, 124, 125m, 130o, 131b, 132, 134o, 136o, 137, 138, 139, 145b, o, 146, 147b, 148, 149, 150m, o, 151, 153, 154o, 155, 156, 157b, m, 159lo, ro, 160o, 161m, o, 162, 163b, 164m, o, 165, 166, 167, 168, 169o, 170, 171, 172b, 173b, o, 174, 175, 176m, o, 177, 178, 179, 180, 181, 183b, 184o, 185, 186, 187o, 188, 189, 190, 191, 192, 193, 194, 195b, 196, 197, 198, 199, 200o, 201, 202, 203, 205m, 206, 207o, 208, 209, 210, 211, 212b, 213b, 214, 215, 216b, o, 217, 219, 220
Collectie Willem H. Moojen: 7, 9, 10, 14, 16b, 19o, 20, 22, 23, 25b, 26o, 29, 30, 31, 32m, 36b, 37b, 38o, 39o, 42o, 43b, o, 46b, 47, 48b, 49o, 51b, 59, 62b, 65o, 70b, 71o, 73b, 75o, 76o, 78o, 82b, 83b, 85o, 87o, 89b, 95o, 96o, 97b, o, 100o, 101o, 102o, 105b, 106o, 107o, 108rb, o, 113b, m, 114, 120b, 131o, 134m, 135, 136b, 141b, 142, 143, 144o, 145m, 152, 154b, 158, 160m, 161b, 176b, 183o, 184b, 205b, 213o
Foto J.G. Nierop: 129b
Foto B.J. Nixon: 92
Foto Jean M. Otten: 93b, 98o, 117b, 163o
Foto J. Pauwels: 133m, 134b,
Foto Photoclair: 86
Foto Tom Rayner: 26b, 50b, 91o
Foto Real Photographs Co: 43m, 64b
Foto C.L. Reynolds: 112o
Foto R. Roozendaal: 195o
Foto Selim San: 119o
Foto Jan Schiet: 68, 69b
Foto Schutte: 53m
Foto F.R. Sherlock: 91m
Foto Signum Ontwerpers Gert Wienbelt: 218
Foto Peter Stewart: 182
Foto Theunissen: 122b
Foto Charles Trail: 47
Foto Waterweg Photos Hans Kraijenbosch: 117o, 125o, 128, 140o, 172o
Foto World Ship Photo Library: 21b, 34, 35, 44b, 50o, 52b, o, 53o, 56b, 58b, 60, 61o, 113o
Collectie H. Zuur: 169b

The author would like to express his sincere appreciation to the various individuals and shipping companies for their cooperation and for the use of photos.

Filip Bruins, Hardinxveld
Fam. van Donderen, Delfzijl
Ben Th. Gernaat, Goutum
Dick Gorter, Castricum
Albert Bazuin, Kolham
Niels Bazuin, Roden
Wessel L. Feenstra, Harlingen
Pieter Pomp, Vogelwaarde
H.J. Smith, Barendrecht
Arklow Shipping Nederland B.V., Rotterdam (Jon Scholten)
BigLift Shipping B.V., Amsterdam (Willem Fokker)
Boomsma Shipping B.V., Sneek
Broere Shipping B.V., Dordrecht (Raoul Baart)
Cebo Marine B.V., IJmuiden
Chr. Kornet & Zonen B.V., Werkendam
EP Shipping & Trading B.V., Rozenburg
Feederlines B.V., Groningen (Rob de Vries)
Flinter Shipping B.V., Barendrecht
Holwerda Shipmanagement B.V., Heerenveen
Jaczon B.V., Scheveningen (Dirk Vink)
Kooy Shipping B.V., Pijnacker (Annette van der Kooy)
Kustvaart Harlingen B.V., Harlingen (Robin Switynk)
Kustvaartbedrijf Moerman B.V., Ridderkerk (Felix Feléus)
Marin Ship Management B.V., Farmsum
Navigia Shipping B.V., Groningen (Dick Jansen)
Onego Shipping & Chartering B.V., Rhoon
Oost-Atlantic Lijn B.V., Pernis (Thomas D. van Gorkom)
Q-Shipping B.L., Rhoon
Sam Shipping B.V., Papendrecht (Gerrit Westerbeek)
Seatrade Groningen B.V., Groningen
Soetermeer Fekkes Cargadoorskantoor B.V., Zwijndrecht
Spliethoff B.V., Amsterdam (Jan van den Bos)
Star Bonaire B.V. (C. Koole Tanktransport B.V.), Zaandam
Theodora Tankers Services B.V., Dordrecht (Arthur Groeneweg)
Wagenborg Shipping B.V., Delfzijl (Henk Zuur)
Wijnne & Barends B.V., Delfzijl